ALSO BY PETER HOTTON

So You Want to Build a House
So You Want to Fix Up an Old House
Coal Comfort

So You Want to Build an Energy-Efficient Addition

So You Want to Build an Energy-Efficient Addition

Peter Hotton

Drawings by Marilynne K. Roach

Little, Brown and Company　　Boston · Toronto

COPYRIGHT © 1983 BY PETER HOTTON

ALL RIGHTS RESERVED. NO PART OF THIS BOOK MAY BE REPRODUCED IN ANY FORM OR BY ANY ELECTRONIC OR MECHANICAL MEANS INCLUDING INFORMATION STORAGE AND RETRIEVAL SYSTEMS WITHOUT PERMISSION IN WRITING FROM THE PUBLISHER, EXCEPT BY A REVIEWER WHO MAY QUOTE BRIEF PASSAGES IN A REVIEW.

FIRST EDITION

LIBRARY OF CONGRESS CATALOGING IN PUBLICATION DATA

Hotton, Peter.
  So you want to build an energy-efficient addition.

  Includes index.
1. Dwellings—Remodeling.   2. Dwellings—Energy conservation.   3. Solar energy.   I. Title.
TH4816.H66   1983   643'.7   83-14879
ISBN 0-316-37385-0 (pbk.)

MV

Published simultaneously in Canada
by Little, Brown & Company (Canada) Limited

PRINTED IN THE UNITED STATES OF AMERICA

To my father,
Nicholas,
who never built an addition
but insulated his house back in 1938

Contents

Introduction   xiii

PART I: WHAT TO BUILD AND HOW TO KEEP IT WARM

Chapter 1   3
Up, Out, Over, or Under: Which Way to Add and How to Plan

Chapter 2   10
The Heart of the Matter: Superinsulation and Solar Heat

Chapter 3   14
Superinsulation: How to Do It, Briefly

PART II: THE NITTY-GRITTY: BUILDING TECHNIQUES

Chapter 4   25
From the Bottom: Excavating and Foundations

Chapter 5   42
It's a Frame-up: Floors

Chapter 6   49
The Uppity Parts: Walls

Chapter 7   61
Top Priority: Roof Framing and Sheathing

Contents

Chapter 8   71
On the Up and Up: Dormers and Second Floors

Chapter 9   82
Ups and Downs: Stairways

Chapter 10   87
Lights Fantastic: Windows and Doors

Chapter 11   95
Icing on the Cake: Outside Trim

Chapter 12   101
Top o' the Mornin': Roofing

Chapter 13   112
The Fun Part: Siding

Chapter 14   121
Fussy, Fussy: Electricity, Plumbing, and Heating

Chapter 15   129
The Smoothies: Walls and Ceilings

Chapter 16   136
Made for Walking: Flooring

Chapter 17   144
The Penultimate: Indoor Trim

Chapter 18   153
The Ultimate: Paint, Stain, Varnish, and Other Good Things, Inside and Out

Chapter 19   160
Before and After: Converting Garages, Breezeways, and Basements

Chapter 20   175
Living on the Outside: Porches, Decks, and Patios

Contents

PART III: DON'T GIVE UP THE HOUSE: SUPERINSULATING AN EXISTING HOUSE

Chapter 21	191

The Easy Parts: Attic Floor and Basement Ceiling

Chapter 22	194

The Hard Part: Superinsulating Existing Walls (and Windows and Doors, Too)

Chapter 23	204

Gasp! Air Exchange and Solutions

PART IV: THE HOME FIRES: SOLAR AND OTHER HEATING

Chapter 24	209

You Are My Sunshine: Solar Devices and Collectors

Chapter 25	215

Just So: Thermal Mass for Heat Storage

Chapter 26	217

Warmth Plus: Auxiliary Heat

Glossary	220
Further Reading	237
Index	238

Introduction

High costs of houses, new or old, and the high cost of money have forced Americans to do two things: build onto their existing houses rather than move to bigger quarters, and do some or all of the work themselves.

Virtually any house can be added to, whether the addition is a wing, a second story or dormer, or a converted garage or basement. This book tells all about adding on, the types of additions that can be built, and, more important, how to build them. It is valuable even if you don't plan to do the work yourself; knowledge of how something goes together will arm you with the tools to deal with architects, contractors, and other professionals who will do the job for you.

Another high cost—that of heating fuel—is making Americans recognize that superinsulation, or at least better insulation than was standard in the past, and the thorough buttoning-up of the house, is a means of saving up to 100 percent of their fuel outlay. At the very least, a well-insulated building will keep the fuel bills at an affordable level.

A superinsulated addition is one thing. It will require considerably less fuel—indeed, almost none if done right, than the house does. So why not superinsulate the existing house? That is trickier, more difficult, and may not be worth the effort or the money, but it is at least worth looking into.

While you're at it, solar heat, in both addition and existing house, can also reduce those fuel outlays. A passive solar heating system is the wave of the future. It takes advantage of the ultimate freebie, the sun; there are no moving parts, nothing to break down, just appropriate devices that harness, retain, and release solar energy.

This book is divided into four convenient parts, to ease the confusion that can arise when considering several things, such as an addition, superinsulation, how to build, and passive solar heating techniques. The first part covers what kind of addition to build and how to get started, and the theory and materials for superinsulation in general; the second, how to do it all; the third, superinsulating an existing house; and the fourth, the solar approach. Any terms used in the book that are not defined in the text will be found in the Glossary.

There may be no doubt that you need an addition. To build it yourself, incorporating in it superinsulation and solar heating devices, all you need is a little talent and a lot of enthusiasm. If you do some or most of the work yourself, you will save up to half the cost.

part I

What to Build and How to Keep It Warm

chapter 1

Up, Out, Over, or Under

Which Way to Add and How to Plan

An addition can take many forms. It can be an attachment to, or a whole new second story on, a ranch house. It can be a dormer, taking advantage of attic space in a two-story Cape Cod-type house, or the third-floor attic of a more traditional building. It can even be a special dormer, one with a half-gable roof replacing a roof that is too shallow or too low for a traditional dormer. It can be a conversion of a garage or of a basement.

Your existing house and plot of land may dictate or restrict your options, but there is probably no house that cannot be added to. If you plan some kind of solar heating system, particularly a passive one, the orientation of the house or the addition is important. Sometimes solar heating is impossible, so other heating systems must be considered. On the other hand, with an energy-efficient addition, with superinsulation and total tightening up, the solar aspect may be unnecessary.

If you have yard room or if you can get a special exception or waiver of any zoning restrictions, an *ell* or *wing* might be the best bet for your addition. It allows the most freedom of planning for size, shape, and number of stories. Some of the possibilities are: a one-story shed-roof ell off a two-story house (among the easiest; Figure 1); a one-story gable-roof ell on a one-story ranch (Figure 2); an extension of the length of the house (Figure 3); a two-story ell on a two-story house (Figure 4). There are many variations.

There are variations on the foundations of such additions, too. Generally, it is not necessary to build a full basement under an ell; basements are not always dry (in fact, traditional basements are not places to store clothes or bedding, even if they seem dry). You could plan an ell to be on the same floor level as the existing house, or be below it. If it is on the same level, a good foundation is one with a crawl space. If it is below the house's floor level (sort of a sunken living room or family room), it could be a slab on grade.

The traditional and most efficient *dormer* is the shed dormer (Figure 5), with a shallow roof extending along one half of the roof, preferably in the back, so as not to disrupt the style and lines of the house. If the house is oriented so that its gable is facing the street, one dormer is going to look awkward. Solution: two dormers, one on each side of the house (Figure 6).

One reason for a shed dormer is the lowness of the ceiling on the second story of a Cape Cod-type house. In a Cape, the roofline comes right down to the top of the first-floor wall, leaving a lot of wasted space where the roof slants down.

Sometimes the roof is quite high, and a shed dormer is not necessary. Then you can build eye, or "A," dormers (Figure 7), which basically pro-

Up, Out, Over, or Under

FIGURE 1. One-story shed-roof ell, or wing, addition. Here the roof comes too close to the second-story windows; windows should be shortened.

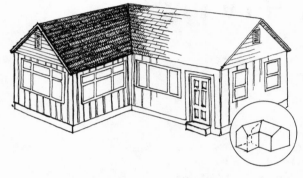

FIGURE 2. One-story gable-roof wing, creating an L-shaped house.

FIGURE 3: Two-and-a-half story wing with dropped roofline.

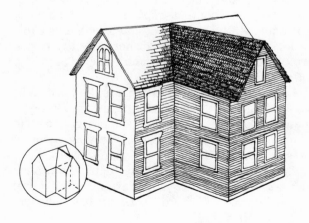

FIGURE 4. Two-and-a-half-story wing, creating a T-shaped house.

FIGURE 5. Shed dormer, added to a second-story roof.

FIGURE 6. Two shed dormers on a third-story roof. Since the house faces the street, only one dormer would look lopsided.

FIGURE 7. Two "A," or eye, dormers.

FIGURE 8. Half-gable dormer, effective where the roof is too low for a standard shed dormer. This is good for passive solar heat if the clerestory windows face south.

vide light and little else except ventilation. Eye dormers are attractive from the outside and interesting from the inside. They are a little tricky to build, particularly with a gable roof going into a slanting roof.

If the roof is too shallow for an ordinary shed dormer, and there is no need or desire to raise the roof to make a full second story, then half the roof can be raised, making a *half-gable* right at the peak (Figure 8). This can work on a one-story ranch, or on a shallow attic above a two-story house. It is particularly useful if the windows below the half-gable face south, giving some solar gain. Of course, the space below the existing roof can be useful for storage.

Raising the roof of a ranch house is another technique, useful if zoning or the size of the yard does not allow an ell. The second story could cover half of a ranch (Figure 9) or the entire house, or could include an overhang, making the added-to house a garrison (Figure 10).

Garages are tempting places to expand to. Today's automobiles being what they are, with rustproofing and other improvements, the need for a garage to house just a car is less important than it was a few short years ago when the battery had to be kept warm. Besides, it would probably be cheaper in the long run to convert the garage (Figure 11) and build a new one, if you or anyone else insisted on it. Of course, a two-car garage is easier to convert, and more practical, than a one-car unit, simply from the standpoint of size. With a one-car unit, you can convert it and add a second story.

Up, Out, Over, or Under

FIGURE 9. *A second floor built on part of a house.*

FIGURE 10. *A second floor added to the entire house, creating a garrison style, with front and back overhangs.*

FIGURE 11. *A converted garage, with French doors and patio; new garage is added.*

A garage conversion may make necessary a connector that is more than a breezeway—although the breezeway could be converted, too. The need for a connector would also apply to a separate building, if that was the plan for the addition.

Converting a *basement* is an obvious and quite inexpensive way to add. Today's raised ranches, or split entries, as they are sometimes called, are particularly adaptable to this type of conversion. And since relatively little of the foundation is below ground level, dampness is reduced or eliminated.

Look Before You Leap

Having considered the choices of addition available, you want to decide which one is most suitable. Planning ahead is essential.

First, ask yourself these questions:

• What is the purpose of the addition? A bedroom? Family room? Dining room? A combination of these? An in-law apartment? Should a bathroom be included?

• How big should it be for a certain purpose?

• What will be the access from the existing house?

• Will there be or should there be an outside entrance?

• If the addition is on the second or third floor, or if it is two stories high, how much room will a stairway take up, and perhaps take away from space in the existing house? What kind of stairway should it be?

• If an addition is on top of an existing structure, is the first floor strong enough to support it?

• How many windows and doors, on each side, will it have—particularly on the south side, if there is one, where there can be solar gain? On the north side, where windows are a liability?

• What kind of foundation? A full basement? A crawl space? A slab on grade?

• Should you mix or match? That is, make an addition the same style as the existing house or make it different? What kind of siding? What kind of roofing?

Only you can answer these questions. Planning ahead will answer them better than anything or anyone. Many of the questions are easy to answer because of your own particular needs and situation. But sometimes questions as to the size of an addition are harder. You don't want to make it too small (people tend to plan smaller than they need) because it would be difficult to add to the addition. You also don't want to make it too big, so that when the family flies the coop you're rattling around in a big barn of a house—and addition.

Draw a tentative plan. Put your ideas and needs and dreams down on paper, where the right and wrong things about them will become more obvious. There are kits and books on the market that allow you to draw rooms to scale, and even cutouts of furniture, also to scale, to allow you to plan a room without overcrowding it.

At this point, or even at the beginning of your planning, consider an architect. One doesn't always cost an arm and a leg, as is popularly thought. Often you can arrange a fee before work is started. Many architects specialize in designing additions and recognize their purposes. Some also do contracting. And some contractors will help you with the planning.

You should also determine legalities. You will need a building permit, which will require a plot plan—where the addition is going to go in relation to the existing house—plus a floor plan, and possibly more specifics. It is not always necessary to have the plans drawn professionally.

You will want to check with your local building department not only about the permits you need but also about zoning and restrictions. Does your addition go within a certain distance from the side borders of your property, for example? There are stipulations as to how close you can build, to both side and front and back borders. If your addition will come closer than is allowed, you might be able to get an exception or a waiver to the law, but this requires hearings and approval of town boards, not to mention neighbors and abutters. These can take time and money, so it might be the better part of valor to work out an addition that does not need an exception or waiver.

Your building department will also tell you exactly what kinds of permits you will need, legally. There are many types: a construction permit that allows you to get started; specialty permits for such things as plumbing, electricity, gas, public works (water and sewer connections), and heating; and so forth. Not all of them may be required. The construction permit usually requires inspections: after the foundation is installed but before backfilling; on framework and sheathing before insulation is installed; and a final inspection that, if you pass, allows you to occupy your new addition. It is called an occupancy permit. Inspections usually go with the specialty permits, too; in the case of electrical work, only a licensed professional can obtain the permit. That doesn't mean he will do the work, but he is responsible for it.

There are three ways to build an addition:

1. Obtain a contractor (it is best to find at least three and let them bid for the job), and let him do all the work and worrying. All you have to do is look at the job and make sure, as best you can, that it is being done to your satisfaction.

2. Act as your own contractor, and hire or invite bids from subcontractors to do various parts of the addition. This entails a lot of headaches, including obtaining insurance, scheduling, buying materials, and keeping subs from tripping over each other while they work.

3. Do the work yourself, except such things as electrical work, plumbing, heating, and sewer connecting, which you either cannot do by law or are not inclined to do.

Your talent, time, and inclination may dictate what you actually do on the addition. But even if you don't do a great deal, reading up on techniques will help you work with contractors intelligently. You will have a working knowledge of the addition and building techniques, and will not be thrown by gobbledygook from a contractor. Most pros are very helpful and really know their business. Some will gladly tell you things; others are reluctant to reveal their secrets for fear you might become a contractor and put them out of business.

One of the most frustrating parts of building is finding the right people: architect, contractor, subcontractors. People don't seem to know where to start and often want some magic formula for finding the best individual at the lowest cost. It doesn't happen that way. It is a matter of perseverance, gut feelings, and stick-to-it-iveness.

Where to start? The Yellow Pages is a good place. And here's a trick: if you can find a Yellow Pages that is ten or so years old and locate architects and contractors who are still in business, you can be confident they have not been run out of town on a rail. Finding old Yellow Pages is another trick; check with the telephone company.

Word of mouth is another good way of finding people. Check with neighbors and people who have had work done. Was their contractor okay? Expensive or not? They should be able to tell you immediately and in no uncertain terms.

Newspaper ads are another good source. The classified section of large Sunday newspapers is sometimes a cornucopia of information and sources. There are many classifications; such categories as "Market Basket" and "Services and Repairs" can come up with contractors, renovators, addition builders, handymen, kitchen experts, and so on. Remember, this is only a source.

Once professionals are located, it's bid time. It is prudent to obtain three bids if possible. Simply tell the professional what you want; if he wants the job he will present a proposal and a cost estimate. If you get three replies (not always easy) and they are reasonably close to one another in cost, you can be pretty sure you are getting a good deal, at least a fair one. If two bids are high and one is low, think twice about that low bid; the bidder may be desperate to get the job and might have cut corners. If the bids are far apart and no two agree, it's time to punt.

Of course, you can avoid some of these situations if you are sure each bidder is bidding on the same things. Architect's or contractor's plans should specify exactly what the addition is going to be built of. A contractor's bid will include materials, so make sure someone isn't substituting an inferior material or smaller piece of lumber than is called for. If you want to buy the materials yourself, you have to be certain that your "lumber list" is accurate; that is, complete enough to finish the job, with allowances for minor shortages or overruns.

Obviously, hiring a general contractor and letting him do all the work will be the most costly way of building an addition, perhaps as high, at current prices, as $70 per square foot of occupied floor space. So if your addition is 20 by 20 feet on two floors, that's 800 square feet; at $70 per square foot, that's $56,000. Don't flinch. It can happen. Perhaps $40 or $50 per square foot would be more reasonable, and at that the addition would cost $32,000 to $40,000. Such costs include everything: foundation, roofing, finishing, absolutely everything. If you build a dormer or raise the roof for a second floor on a one-story ranch house, the cost per square foot will be less, because you don't need to excavate or build a foundation.

How much can you save by being your own contractor? Maybe 15 percent. The saving for doing it yourself is about 50 percent. If there is a reasonable amount of electrical, plumbing, and other work for professionals to do, the saving will be about 40 percent.

Under any circumstances, whether you contract, subcontract, or do it yourself, be sure to get a detailed copy of your local building codes, so that you follow all the rules. Some of the places where building codes may specify materials or dimensions will be pointed out in the how-to-do-it chapters.

You are ready to go—plans made, needs determined, architects and contractors ready to be hired. But first, how about designing your addition around superinsulation and the use of solar energy, so that you can conserve the fuels the modern world is finding in shorter and shorter supply, and save on those astronomical heating bills that have plagued us all in the past few years?

chapter 2

The Heart of the Matter

Superinsulation and Solar Heat

Superinsulation of a house or an addition is no different from standard insulation; there is just more of it. Vapor barriers play an essential part, as does ventilation. The sole purpose of superinsulation is to conserve heat, to reduce or even eliminate heat loss through the walls, ceilings, and floors. This is accomplished by wrapping up the house or addition so well as to make it as nearly airtight as possible, for it is through air exchange that a house loses heat. An insulated house that has an air exchange of once, twice, or more often per hour isn't holding heat very well. A well-sealed, superinsulated house will have an air exchange of once every two hours or even less.

Not so long ago insulation and heat moguls were saying that it was possible to overinsulate a house; that extra insulation was too expensive to make it worthwhile. With high heating costs, those days are gone forever. Now it is not possible to overinsulate; if you can build a house or addition that loses no heat, you have the ultimate. Once you heat the interior space to, say, 70 degrees, it will stay that way, barring the opening and closing of doors to go into and out of the living space.

Insulation is essentially dead air. An uninsulated house has an air space in its wall cavities. If this air were dead—that is, not moving—it would be an excellent insulator. But because many chinks and cracks and holes allow the air to move, it is not dead and therefore doesn't insulate. No one has come up with a proposal to make a hollow wall airtight, and so insulation was invented. It usually consists of a material that creates many small cells of trapped air, air that remains still and thus insulates.

TYPES AND EFFECTIVENESS

Insulating materials are fiberglass, mineral wool, cellulose, Styrofoam (polystyrene foam), urethane foam, polyisocyanurate, and other, less well-known, not commonly available materials.

These substances come in different forms: as batts or rolls of material prefabricated into certain thicknesses (fiberglass), and set on attic floors, squeezed or stapled into walls, and hung on ceilings; as poured or blown material (fiberglass, mineral wool, cellulose); and as rigid material made into boards (Styrofoam, urethane, polyisocyanurate). This last material may be somewhat unfamiliar: it is a foam board covered with aluminum foil on both sides, and has double the insulating value of Styrofoam. It is known by its brand names: High R Sheathing, sold by Owens-

Corning; Thermax, made by Celotex; and others.

The effectiveness of insulation is measured in R factors—a term to express the resistance of a material to the loss of heat from a heated area to a colder one. The higher the R number, the greater the resistance. A standard insulated wall has an R factor of 11, the result of 3½ inches of fiberglass in the wall cavity. Other materials on the walls, such as wood sheathing and siding on the outside and plasterboard or plaster on the inside, add another 3 to the R factor, bringing the total R factor to 14. A standard ceiling (attic floor) has an R factor of 19, with 6 inches of fiberglass.

With superinsulation, here are the R factors attainable through the use of more insulation (fiberglass):

Basement ceiling	R 19	6 inches
Walls	R 38	12 inches
Attic floor	R 40–57	12 to 18 inches

Even a cathedral ceiling, which is difficult to superinsulate properly, can, with 6 inches of fiberglass plus rigid insulation, attain an R factor of 27 to 32.

Fiberglass is among the most commonly used materials. It may also be the safest because it is basically fireproof. Mineral wool is, too. The other materials may pose a fire hazard. While cellulose has been treated with chemicals to make it fire-resistant (in fact, it must be rated Class A in order for it to be used), there are questions as to how long it will remain so, particularly if it becomes damp or wet. Styrofoam is flammable, but its fumes are not considered toxic. Urethane and polyisocyanurate are flammable and their fumes are toxic. (What constitutes toxicity may be confusing because fumes from burning wood are also toxic.)

Superinsulation, then, requires super-thickness. Modern science has not yet come up with a practical, inexpensive, fire-resistant insulation that is less than several inches thick. Until it does we must stick with thickness.

A second, essential aspect of superinsulation is the appropriate vapor barrier and air stop (one and the same). The name tells all: the vapor barrier stops the passage of water vapor from the inside of the building into or through the walls, where its condensation might lead to mildew and eventual rot or decay of wood members; and it also stops the passage of air. The stoppage of air is the retention of heat. The vapor barrier is always placed against warm (inner) surfaces, not cool, outer ones. With a full vapor barrier, and with all cracks and crevices well sealed, the house or addition becomes essentially airtight.

Vapor-barrier material is normally a layer of aluminum foil or kraft paper, or—the best—6-mil polyethylene plastic, a semiclear material that comes in rolls. It is better than foil or paper because it will hold up longer and because it can be applied in a continuous sheet, over joists and studs as well as insulation. Vapor barriers that are part of rolls or batts obviously will leave joists and studs unprotected. All these materials— aluminum foil, kraft paper, and polyethylene— are flammable and so must in turn be covered with a nonflammable material.

In making a building airtight it is also important to seal cracks and holes with caulking during construction. Electrical wires, plumbing pipes, and other ducts and tubes can be located so that they do not penetrate walls or ceilings; they can come underground into the cellar and then up through the walls—we shall see how—into the building. It is important to plan for these pathways in building a superinsulated addition.

The Heart of the Matter

The Need for Ventilation

When a vapor barrier envelops the whole house or an addition, water vapor can build up *inside* and can condense on cool surfaces, such as windows and other glass areas, or in restricted places where it can cause mildew and decay. Also, odors, smoke, gases, even germs, can pollute the air within a house. Obviously, then, ventilation becomes a necessity.

One way to ventilate is to open doors and windows twice a day—but this defeats the purpose of superinsulating to save heat. Another possibility is to use exhaust fans, but they may be inadequate and may also pull heat out of the house. If only an addition is superinsulated and airtight and the rest of the house is not, probably there won't be too much of a water-vapor buildup and ventilation will take care of itself. On the other hand, if you superinsulate not only the addition but also the existing house, then ventilation is a must: in the attic, to prevent buildup of heat in the summer and water vapor in cold weather; in basements and crawl spaces; and in insulated roofs. The walls in a superinsulated house are built to provide the ventilation they need (see Chapter 6).

You may, however, want to consider an air-to-air heat exchanger, to permit the necessary ventilation and at the same time conserve the precious heat. We will have more to say about these artful devices in Chapter 23.

FIGURE 12. Sun space, not only for growing plants (a greenhouse) but also for heating the house.

Solar Heat

Closely related to superinsulating, because it is another means of conserving on fuel, is passive solar heating. By careful determination of the

Superinsulation and Solar Heat

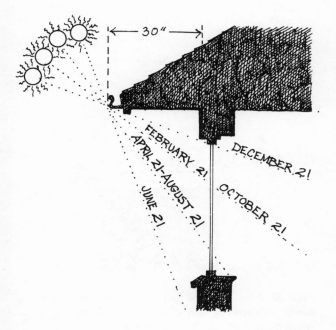

FIGURE 13. *A 30-inch overhang is necessary to allow winter sun to enter house for passive solar heating and to prevent hot summer sun from entering.*

location and orientation of your addition, you can heat it with the free energy of the sun. A large part of the addition must face south, or at least within 10 to 30 degrees of south; at least 10 percent of the wall area must be devoted to windows; and 70 to 85 percent of those windows should face south.

You can also include sun spaces, greenhouses, and skylights to increase the amount of entering sunshine (Figure 12). The glass used in these elements must slant properly to catch the sun's rays at the most direct angle. It is important, too, to make sure that the solar areas can be shaded to stop the sun's heat when it isn't wanted. The eaves of your addition, for example, should be wide enough to shade a south wall from the summer sun but not so wide as to stop the winter sun (Figure 13). A way to do this is to make the eave overhang 30 inches wide and locate the tops of windows about 16 inches below the ceiling. Glass areas should also be ventilated to allow the escape of overheated air in hot weather.

There are many other ways to take advantage of solar heat. Masonry and stone, to store heat, can be used in various parts of the addition. Barrels of water, painted black, are effective solar heat-holders in a greenhouse or sun space. More about solar heating and storage is in Chapters 24 and 25.

If an addition is superinsulated, airtight, and heated with solar energy, there is no need for heating further, and little need to cool it except in extremely hot climates or weather. The present heating system of the existing house can be extended if it is large enough, but it probably won't be necessary. To determine exactly how much heat you get and how much more you might need, you may want to live in the addition for a year and then make a well-educated decision.

chapter 3

Superinsulation

How to Do It, Briefly

Superinsulation is actually quite new in most construction; there are no standard materials or techniques. In this chapter we give you an overview of the places in building an addition where superinsulation is particularly appropriate and a general description of some of the techniques you can use. The details of actually putting it in and providing for the needed ventilation are found in the chapters that follow.

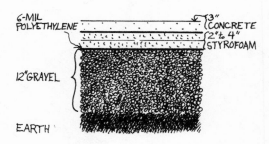

FIGURE 14. *Two to 4 inches of Styrofoam insulation under a basement slab, with a vapor barrier of polyethylene just above the crushed stone.*

FOUNDATIONS

Foundations for basements or crawl spaces have the least insulation of any part of a superinsulated addition because most of the foundation is underground. Thanks to the intrinsic heat of the earth below three feet down, only a little insulation is needed. Generally, 2 to 4 inches of Styrofoam are placed under the basement slab (Figure 14), on top of crushed stone and a vapor barrier. Two inches of Styrofoam are glued to the outside of the foundation (Figure 15), with construction adhesive or one specified for use with polystyrene. Wherever Styrofoam is exposed, outside or in, it must be protected against wear and tear and also must be covered because it is flammable.

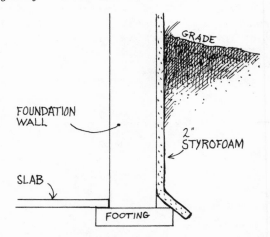

FIGURE 15. *Two-inch Styrofoam glued to the foundation; slanting the insulation guides water away from the footing.*

How to Do It, Briefly

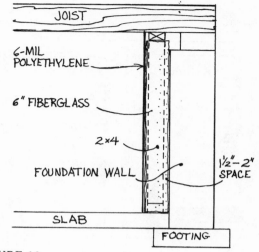

FIGURE 16

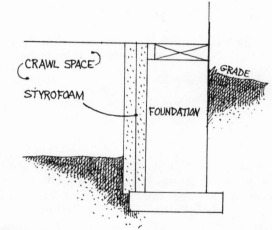

FIGURE 17

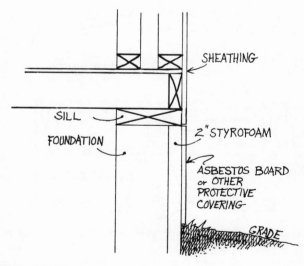

FIGURE 18

Another technique in this area is to install fiberglass in a stud wall on the inside of the foundation (Figure 16). This wall can be built of 2 x 4s and set 2 inches in from the foundation wall, which will permit filling the wall and the space with 6-inch batts of fiberglass. If the basement is to be living space and is heated, this type of insulating is particularly practical. Styrofoam can be added to this fiberglass wall, too. The foundation treatment for a crawl space is similar (Figure 17).

With a concrete slab on grade, 2 to 4 inches of Styrofoam are applied directly under the concrete and on top of a vapor barrier. Since the slab needs a foundation, it too must be insulated; in this case the insulation can only go on the outside. When any thickness is put on the outside, the frame wall should extend beyond the foundation so that it is flush with the insulation (Figure 18). Sheathing and siding provide the necessary overhang beyond any protective material covering the Styrofoam.

FIGURE 16. Six-inch fiberglass in a 2 x 4 stud wall set slightly away from the foundation.

FIGURE 17. Two to 4 inches of Styrofoam glued on the inside of a crawl-space foundation.

FIGURE 18. With insulation on the outside of the foundation, the frame wall must extend beyond the foundation to prevent the insulation from creating a rain-catching lip. The siding will cover the gap between insulation and sheathing.

Superinsulation

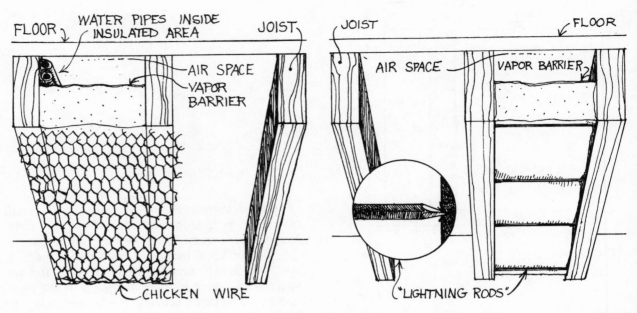

FIGURE 19 A. *Fiberglass insulation in a basement or crawl-space ceiling, held in place with chicken wire. Water pipes are between the floor and the insulation.*

FIGURE 19 B. *Pointed steel wires, called "lightning rods," are stuck between the joists to hold up insulation in a ceiling.*

Floors

There has always been a controversy about the insulation of the first floor of a traditional house or addition. One school of thought maintains that the heat from the furnace will rise and penetrate the uninsulated floor, thus warming the first floor. The other school notes that an unheated basement or crawl space is colder than the occupied, heated space above it; therefore there will be heat loss through the floor if it isn't insulated.

So it is best to insulate. Generally, 6 inches of fiberglass with vapor barrier up are installed between the joists and kept in place with stapled chicken wire (Figure 19 A) or "lightning rods" (pointed wires jammed between the joists; Figure 19 B) or a ceiling. It is also important to install insulation along the perimeter of a basement or crawl space at the sill (Figure 20). More about putting in floor insulation in Chapter 5.

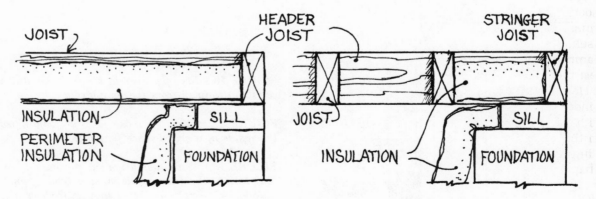

FIGURE 20. *Insulation around perimeter of basement or crawl space, where the joists are at right angles to the sill (left) and parallel to the sill (right).*

How to Do It, Briefly

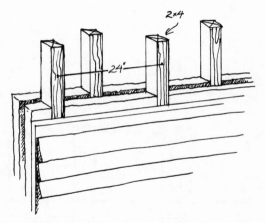

FIGURE 21. *A double wall of studs set on 24-inch centers, offset 12 inches laterally, so that there is no contact between the walls.*

WALLS

Superinsulation focuses chiefly on walls, since they are the part of the house most vulnerable to penetration by cold. They have windows and doors and often cracks and chinks. So it is especially important to make them snug. This is best done by building double walls that can hold up to 12 inches of insulation. The walls can be made of 2 x 4 or 2 x 6 studs. There are variations in the technique, but they are small.

The basic technique is to make two walls, each on its own floor plate and topped by its own top plate (Figure 21). The studs are set on 24-inch centers and those in one wall are offset from those in the second, to prevent any wood members from acting as conduits for heat passage. Six-inch unbacked fiberglass is inserted in each wall (Figure 22) and a vapor barrier of 6-mil polyethylene is stapled on the inner face of the wall with a one-foot overlap at the floor, ceiling, and corners (Figure 23). Or a one-inch-thick sheet of polyisocyanurate can be attached to the inside of the inner wall (Figure 24). This is a vapor barrier by itself, so it needs no further one. But it is flammable and thus must be covered with a fire-resistant material, like plasterboard, on the inside.

Headers, beams, and structural members above windows and doors in a doubled wall are filled with fiberglass or with polyisocyanurate. Details on the treatment of these elements and on handling the floor and sill for a double wall are in Chapters 5 and 6.

FIGURE 24. *One-inch rigid insulation on the inner face of the inner wall will add insulation value. With Styrofoam as rigid insulation install a polyethylene vapor barrier. High R Sheathing has an aluminum foil face, which is a vapor barrier. Seal all seams with tape.*

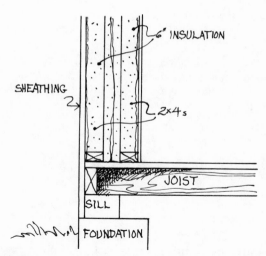

FIGURE 22. *Six-inch unbacked fiberglass insulation is inserted in each unit of a doubled 2 x 4 wall. The spacing of walls 4 inches apart allows insulation units to touch lightly.*

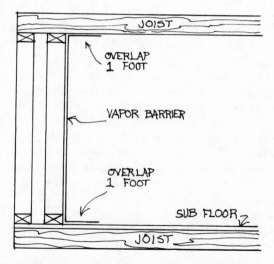

FIGURE 23. *Polyethylene vapor barrier must overlap at ceiling, floor, and corners by one foot.*

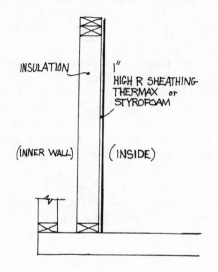

Superinsulation

FIGURE 25. *Insulation between joists, up to the top or near the top of the joists of an attic floor. Another layer of insulation is put on top of this, at right angles to the joists.*

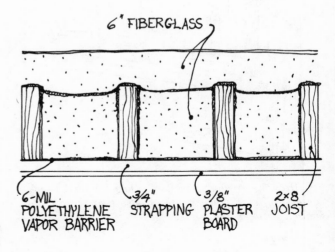

Ceilings

Ceilings are the easiest part of a house or addition to insulate, particularly when they are under construction. They are also important, for since heat rises, a lot of heat is lost through the top of a house.

For the ceiling immediately below an attic (which can be thought of as the attic floor) it is best to put a polyethylene vapor barrier under the joists, then to put up the ceiling material. With this in place, drop in 12 inches of fiberglass (Figure 25). The ceiling will keep it from sagging.

It is not, however, really as simple as "dropping." Twelve inches of fiberglass will extend above the tops of the joists (Figure 26). If the joists are 2 x 6s, you can install 6 inches of unbacked fiberglass between them and then put 6 more inches on top, at right angles to the joists. If the joists are 2 x 8s, use two 3½-inch layers of fiberglass between the joists and a 6-inch layer on top, again at right angles (Figure 27). The important point is that there must be at least 12 inches of insulation. Any less simply won't stop the passage of heat; any more won't hurt a bit.

If you want to put floorboards on a portion of the attic floor, install extra joists where the floorboards are to go, on top of and at right angles to, the first joists (Figure 28).

Keep in mind that insulation should not fill the overhang of the roof at the eaves. It can overlap the walls but should be cut before it fills the overhang. If you use pouring insulation, like fiberglass or mineral wool, wood baffles should be installed between the joists to act as dams that keep the loose stuff away from the overhang (Figure 29). Serious condensation of moisture can result if the insulation fills the eaves.

Why not, by the way, use pouring insulation in the attic floor-ceiling? It tends to scatter and

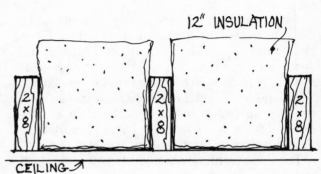

FIGURE 26. *Twelve-inch insulation would extend above 2 x 8 joists in an attic floor, creating gaps where the joists are. This is to be avoided.*

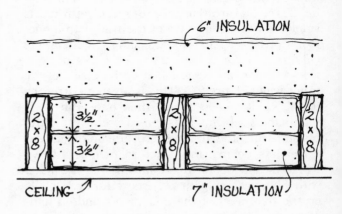

FIGURE 27. *To avoid gaps in insulation, adjust the thicknesses of insulation so that the first layer or two comes to the top of the joists; then lay the next layer at right angles to the joists. This covers all gaps as well as the joists themselves.*

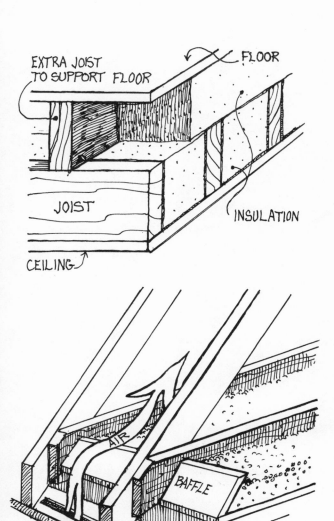

FIGURE 28. *If there is a need for a floor in the attic, extra joists are set at right angles to the original joists and extra insulation is put between them.*

FIGURE 29. *If insulation in an attic floor is the poured kind, wood baffles should be permanently installed to keep the insulation out of the overhang, whether or not it is ventilated.*

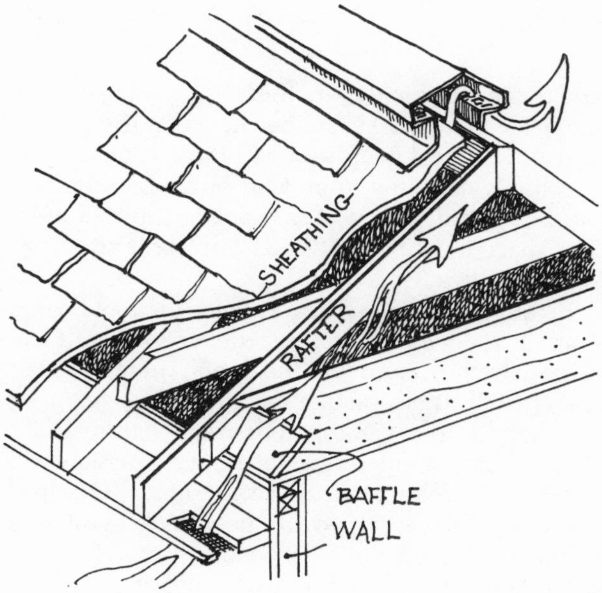

FIGURE 30. *Insulation must not go into the overhang of the roof. Baffles hold it back; soffit vent and ridge vent allow good ventilation.*

generally has a lower R factor than batt or rolled material. Cellulose is not recommended even though it is a good insulator because it is organic and there is no guarantee of its long-range fire-resistance.

Ventilation in an attic is particularly important; the best way in a traditional roof is to have soffit vents and a ridge vent (Figure 30). There is more detail about ventilation in Chapters 7 and 11.

What about a cathedral ceiling? Superinsulation here is difficult to achieve, especially if you expose the beams.

Consider, first of all, *not* exposing them. If the roof joists or rafters are 2 x 12s (somewhat unlikely), you can install 9 to 9½ inches of fiberglass between them and cover it with a ceiling. This thickness would give an R factor of 30 — not superinsulation but almost. With rafters of this size

Superinsulation

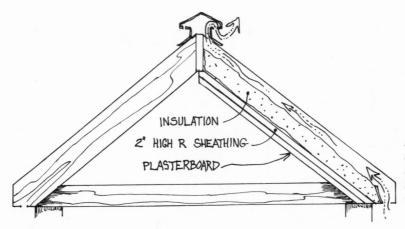

FIGURE 31. *When insulation is inserted between rafters, there is an air space between the insulation and the roof boards. Soffit and ridge ventilation permits the air in the space to move.*

and this much insulation, there is a good result: an air space of about 1½ inches between the insulation and the roof boards (sheathing). That space is absolutely essential to prevent buildup and possible condensation of water vapor in the roof area. The space must be ventilated with soffit and ridge vents (Figure 31).

If you have exposed beams, with the roof boards (called a "deck") forming the ceiling surface, obviously you cannot put insulation there. But you can install rigid insulation (urethane foam or polyisocyanurate) on top of the roof boards, under the shingles. Four inches of rigid insulation will give an R factor of nearly 32 (Figure 32). It is sometimes necessary with such thick insulation to use extra-long nails for securing the shingles. Even better is to build an extra roof of sleepers a full 4 inches wide and 1½ inches thick on top of the sheathing boards. Insulation can go between these sleepers and a plywood roof on top of that (Figure 33). In this treatment a vapor barrier and ventilation are not required.

We must point out, however, that in general a cathedral ceiling with exposed beams is not practical in superinsulation.

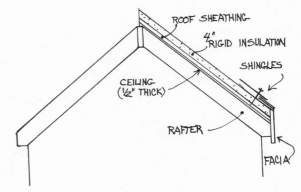

FIGURE 32. *Rigid insulation is applied over the roof in a cathedral ceiling with exposed rafters. Roof shingles are then attached with extra-long nails to pierce the insulation.*

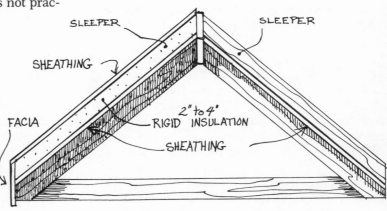

FIGURE 33. *A better way to install rigid insulation on top of a roof is to build an extra roof with sleepers, insert insulation between them and the first roof, and cover with plywood roof sheathing. Shingles are then applied in the normal manner.*

Windows and Doors

The most important thing about windows and doors in a superinsulated building is that they be made to resist heat loss. For windows, this can be done by adding layers of glass (double-glazing, or so-called insulating glass), even three or four layers. You can add fancy windows or outside and inside storm windows. But the more layers of glass, the less light you admit to the house; and if several of the layers are detachable, you have the problem of storage and maintenance.

Thermal shutters may be helpful. They are opaque and thus would be closed at night only—but of course that is the colder time of day in the winter. They can provide an R factor of 3 to 10 in addition to the R factor of the glazing. The simplest shutter is a frame of 1 x 2s butt-jointed at the corners, fastened with corrugated fasteners or finishing nails, and covered with Thermoply (a thin insulating sheathing) or other material. The dead-air space inside the frame is itself an insulator, or you can fill the space with rigid insulation (Figure 34).

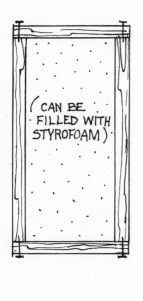
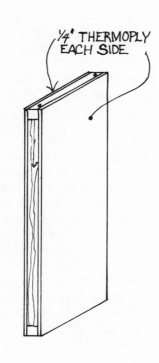

FIGURE 34. *A simple shutter can be made of 1 x 1s or 1 x 2s faced on both sides with Thermoply, a thin insulating material. Hardboard could be substituted.*

Superinsulation

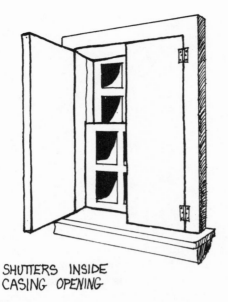

SHUTTERS INSIDE
CASING OPENING

FIGURE 35. *Thermal shutters installed inside the window casing.*

SHUTTERS OVERLAP
CASING OPENING

FIGURE 36. *Thermal shutters installed outside the window casing.*

Shutters can be installed inside the inner window casing (Figure 35) or against the face of the casing (Figure 36). In Chapter 10, there is more information about installing and using shutters in connection with the extra-wide jambs you need when walls are doubled.

Exterior doors are potential losers of a lot of heat. Modern doors are steel-clad, filled with good insulation, and the best ones have an R factor of up to 19. They are also well weatherstripped.

In severe climates where snow and ice can raise havoc, you should avoid sliding glass doors. If you must have them, consider shuttering them or putting inexpensive storm sliders outside them. Good substitutes for sliding doors are French doors—popular what seems like a million years ago, now returning to fashion. These are a series of doors that are hinged and open in the standard way, with large, usually double-paned, expanses of glass. They do need storms or shutters for protection against heat loss.

A Sound Investment

There is another aspect of superinsulation to keep in mind. A superinsulated building is very well protected against the penetration of sound from outside in. If you have noisy children in neighboring yards or a lot of traffic on your or nearby streets, this is an important point. Sound is vibration and travels through the solid parts of a wall. The doubled walls of a superinsulated building are separate; therefore, the vibration is stopped at the gap between the walls. Presto! You are in a quiet haven.

Within the house or addition, good sound control between rooms can be obtained by caulking all joints after plasterboard is put up. Sound-absorbing board can also be put up on the studs before the plasterboard is attached.

Ceilings under attic floors do not have to be treated for sound control because of the great amount of insulation above them and also because attics are not especially noisy. But for other ceilings, rugs and carpet pads will do a lot for reducing impact and other noise.

Finally, in considering the possibility and practicality of superinsulation, what about expense? Are the extra amounts of insulation and the doubled walls going to cost more, and how much more? Some of the expense is reduced by changing traditional spacing of joists and studs, as we will see. And elimination of a central heating system, or the extension of the one already in the house, will compensate for any additional cost. In the long run the savings in heat will pay for the superinsulation.

part **II**

The Nitty-Gritty

Building Techniques

chapter 4

From the Bottom

Excavating and Foundations

Any addition that is not a dormer, raised roof, second floor, or converted basement—in other words, an ell or wing addition—will need a foundation. The foundation of such an addition will not be tied into the foundation of the existing house; it will sit by itself. Only the superstructure of the ell—joists, walls, ceiling, roof—will be attached. Building a foundation can be a do-it-yourself job, but in general it's a good idea to leave the excavating, making of forms, and pouring of concrete to the pros.

First you have to decide just what kind of foundation to build. There are several types: a full basement, a crawl space with foundation, a crawl space with piers, a slab on grade with foundation.

If your existing house already has a basement, chances are good that you don't need another. Moreover, if there is a water problem, a basement, no matter how well built or waterproofed, can be plagued with water or dampness for its lifetime. A basement is damp essentially because of condensation of moisture on cool surfaces, and thus is not a good place to store such things as clothing, paper products, or metal. Proper ventilation and/or a dehumidifier can solve the dampness problem, but a dehumidifier requires electricity, and you probably will want to avoid any additional use of electricity.

Some arguments for a basement: it can provide room for living or some storage space. If the property slopes down, away from the house, the basement could be a walk-in type, where part is level, or nearly so, with the ground. Most of it could thus be above ground and therefore drier and usable as living space. Also, a basement makes a house warmer; but since the basement ceiling and probably other parts are going to be heavily insulated, this argument is not valid.

Probably a crawl space with foundation is the best way to go. An open crawl space, with a series of piers, is sturdy enough, but the open area may be a haven for vermin and other undesirables, and the best material for skirting it is probably concrete. A slab on grade poses insulation problems; even when it's well insulated, not only is a concrete slab hard but its structure makes it inflexible to design things around.

But we will tackle all types.

BASEMENT WITH FOUNDATION

The size and shape of the addition dictate the size and shape of the foundation, no matter what. Unless it is deliberately set at an angle from the house, it should be oriented square to the house, and a guide that is square and level must be made for the excavator. Don't try excavating by hand or

From the Bottom

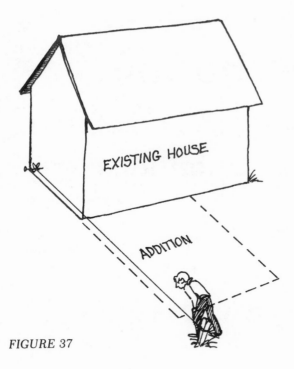

FIGURE 37

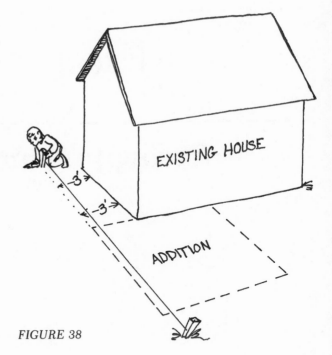

FIGURE 38

by yourself; this is a job for a backhoe, with an expert operator.

Locate your addition and then get it square to the house in this way: if the edge of the addition is to line up with the edge of the house, tie a surveyor's string to a nail driven into the side of the house at the side opposite where the addition is to go. Bring the string along the side of the house, barely touching it, and extend it beyond the house and beyond the length of the addition (Figure 37). Keep the string tight and line it up by sight. This can be tricky; it is easy to move the string away from or against the side of the house. If you line up the string several times, you will get the hang of it. Try it; drive a stake at one spot where you have lined up the string. Then resight down the string and drive another stake a few feet beyond it. Resight and drive a third stake farther down the string. When three stakes line up, you can be reasonably sure that the addition will be square to the house.

If the addition sticks out beyond the house, or butts up against the house a few feet in from its edge, sight the string and measure from it at the edge of the house to where the addition will go (Figure 38, away from the house; Figure 39, in from the house). Use the same measurement at the far end of the string; that is, at the farthest extension of the addition.

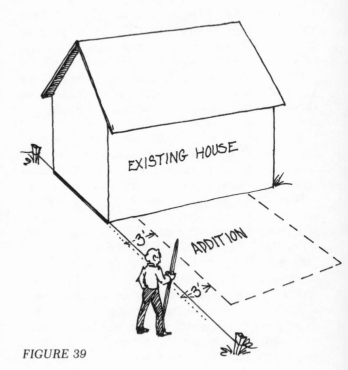

FIGURE 39

Excavating and Foundations

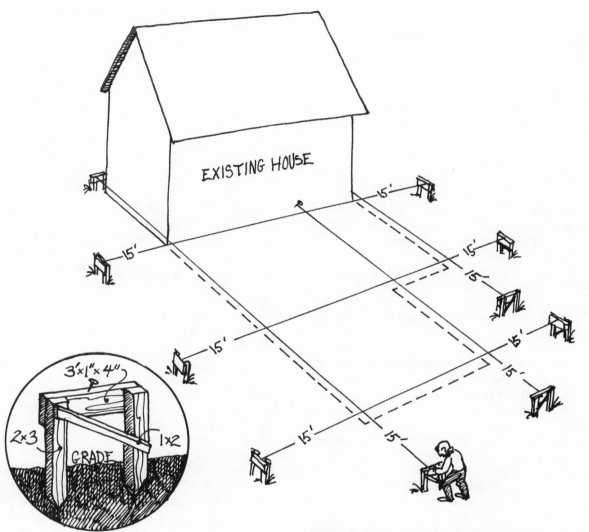

FIGURE 40. *Use batter boards to get corners square when laying out the addition before excavating.*

FACING PAGE:

FIGURE 37. *To line up the addition with the house, run a string along the house wall and extend it beyond the addition's proposed end wall.*

FIGURE 38. *If the addition extends beyond the house, measure out the proper distance from the wall before lining up the string.*

FIGURE 39. *If the addition is indented from the end wall of the house, measure the indentation before lining up the string.*

Once you have the addition marked out square to the house, you must make sure that it has square corners. A carpenter's square is too small to make accurate corners here. You can use a 4 x 8-foot sheet of plywood, the corners of which are square; or you can make a large right triangle of 1 x 3 boards, with one leg 3 feet long and another, at right angles to the first, 4 feet long. The diagonal piece connecting the two legs (the hypotenuse of your right triangle) will be 5 feet long. Another way works whether the addition is square or rectangular in shape: measure the diagonals of

From the Bottom

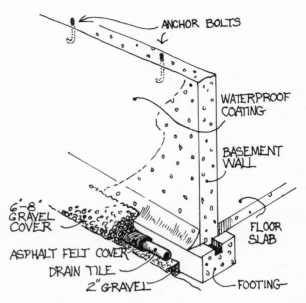

FIGURE 41. *The parts of a foundation. The key in the footing is made by inserting a 2 x 4 into the concrete before it sets; it keeps the foundation from moving laterally.*

the rectangle (from one corner to the diagonally opposite corner). When the two diagonals are equal, all four corners are square—that is, 90 degrees.

Now you are ready to mark the space for digging. Make several batter boards (Figure 40) by connecting a 3-foot 1 x 4 to two 2 x 3 stakes, with a 1 x 2 brace. Six are needed for a rectangle, more if there are jigs and jogs or other ells in the foundation.

Extend the lines of each dimension of the rectangle at least 15 feet, and drive the batter boards into the ground at each extension (Figure 40). Secure surveyor's string to the batter boards so that the strings cross at each corner. The string can be adjusted to make the corners square and can be wrapped around the board's crosspiece or around a nail driven into it.

To make sure the strings cross at each corner, drop a plumb bob (weighted string) from the surveyor's string so that it lines up with the corner. Don't make the weight too heavy or the string will sag.

Accurate location of the foundation is the most important part of the entire project, because from the foundation all other parts of the addition come. If the foundation is not accurate, the addition won't be, even though you will be able to fudge a little when putting on the floor structure and even the walls. But you can't fudge too much.

How high should the foundation be? That depends on several factors. If the addition is to be on the same level as the house, you have to locate that level and determine what kind of sill plate will go on top of the foundation, how deep the joists will be, and how thick you need to make the floor.

The easiest way to do this is to take off the siding and sheathing of the house; you are going to have to take off the siding anyway, where the addition butts up against the house proper. Exposed, the wall will reveal the level of the floor, because the wall is set on the subfloor. Even if it isn't, as in post and beam construction, you will be able to see the floor level and to make calculations by sight. Line up the addition foundation's top with the top of the existing foundation. Then keep all strings level.

How deep should the excavation be? The depth is determined by the height of the foundation walls. The standard height is 7 feet 4 inches from the top of the footings (the concrete pads that hold up the foundation; see Figure 41) to the bottom of the first-floor joists; 8 feet is better. The basement floor is usually 3½ inches thick and is poured on top of the footings, so the actual height of the basement ceiling is reduced by that much.

If the land is level, the work is simplified. A foundation should extend 18 inches to 2 feet above the final earth line in order to protect against termites and to prevent water problems with wood members. So the foundation should extend 6 feet into the ground; less if the earth is banked up on the foundation to allow drainage away from the addition.

The excavation is dug to a uniform depth, and is made level. Then a trench is dug around the perimeter for the footings, which must be placed on undisturbed earth to prevent compressing of

Excavating and Foundations

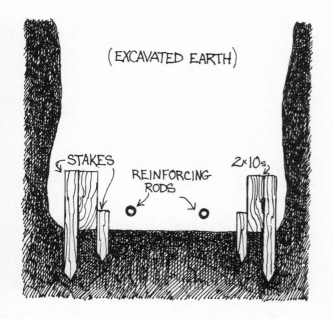

FIGURE 42. *Form boards for a foundation footing. Small stakes inside the forms will not interfere with the footing's strength; it is not necessary to remove them. Another method to keep boards from spreading is to nail 1 x 2s across them.*

soft, backfilled earth and possible collapse of the footings.

Better yet is to dig the entire hole to where the bottom of the footings will go and fill the floor area with 6 to 18 inches of crushed stone. This allows good drainage of water under the basement floor. The footings, which hold up the foundation, must be below the frost line, which is the maximum depth at which soil freezes. This can be 3, 4, or even 5 feet in cold climates. Your building code will tell you.

Footings are as deep as the foundation wall is wide, and twice as wide as the foundation thickness. So if you plan a 10-inch foundation wall, the footings should be 10 inches deep and 20 inches wide.

Form boards for footings can be 2 x 10s. They are secured by stakes inside and out (Figure 42) and must be level along both their width and their length. Use a mason's level (a 4-footer) for measuring all aspects of level and plumb; it is accurate. In fact, the longer the level, the more accurate it is. Oil the form boards to prevent concrete from sticking to them.

Half-inch steel rebars (reinforcing rods) will add strength to the footings. Put two in each footing, halfway up and 2 or 3 inches in from each side. They can be bent to turn a corner, and are a must where a footing crosses a trench for a pipe. Rebars are placed in the footing space and supported by stones so they are midway in the depth space. This is to prevent their moving when the concrete is poured. Once the concrete has cured (hardened), the bars will stay in place.

Stepped footings are built when the land slopes too steeply to allow a full level run. Each run of footing must be level. Remember that the height of each step down or up should be no more than the thickness of the adjacent footings.

Ready for Pouring

It is best to order ready-mix concrete for footings, but be aware that the usual minimum order of ready-mix concrete by the truckload is 3 cubic yards (27 cubic feet in a cubic yard, called just a yard). So if your footings take less than 3 yards of concrete, you will be paying for the 3 yards anyway. You don't have to take it all, and if you haven't any use for it, there's no point in letting it set into an immovable pile in your yard. Chances are good, however, that your footings will take close to the minimum 3 yards.

If you mix your own, use the classic formula of 1–2–3: 1 part portland cement, 2 parts sand, and 3 parts aggregate (gravel or crushed stone). There are many formulas for many uses, but the 1–2–3 mix is strong and simple to keep track of when you mix. You can rent a gasoline or electric mixer for small amounts. Mixing anything more than 2 yards will be a long, hard, tedious job.

When the concrete is poured into the forms, puddle it: that is, drive the blade of a spade or shovel into the concrete, up and down, sort of shaking it so it will settle into all nooks and crannies of the form. When the concrete reaches the top of the form, set a screed, a 2 x 6 or similarly sized board, on each edge of the form and with a helper drag it back and forth (Figure

From the Bottom

FIGURE 43. *Screeding concrete to smooth it off in its forms.*

43), smoothing the concrete out level with the form tops.

Now install a key in the footing by inserting an oiled 2 x 4 into its center along the length before the concrete sets and removing it after the concrete has set. This creates a groove in the footing into which the foundation, when it is poured, will fit (be keyed) so that it will not move sideways (see Figure 41).

Concrete begins to set in 15 minutes, and takes 28 days to cure to its greatest strength. It is strong enough to permit removal of the forms and to work with in a day or two. It cures best in cool, wet weather. In very hot weather and in dry weather, put straw or newspapers on top of it and keep it wet for at least a week or two, or wet it down and cover with canvas. This will slow down the curing and make the concrete stronger. The 2 x 4 key can be removed after a day. Make sure it is greased so that it will come out easily and not take chunks of concrete with it.

At this point, install drain tiles outside of and next to the footings, whether there is a drainage problem or not (see Figure 41). The drain tile picks up underground water and guides it toward a dry well (if the soil is permeable enough to allow drainage); into a sump to be pumped directly to a storm sewer, if any; or as far away from the foundation as possible. Drain tile is pitched slightly toward the dry well, sump, or sewer. It is made of clay or concrete, 4 inches in diameter and 12 inches long, placed on 2 inches or more of gravel and spaced ⅛ inch apart. The top of the tile joint is covered with roofing felt and the tile is covered with 6 to 8 inches of gravel. You can also use plastic or asphalt and fiber pipe with holes in it, covering the pipe in the same manner as the tile.

Remember, when tiles are installed, they must drain *to* someplace to be effective. If they went around the addition and just sat there, several feet below grade, water entering them would also just sit there. So the drainage must be to an area lower than the level of the pipes. If this is impossible, they should drain into a sump so that the collected water can be pumped away.

When turning a corner of the foundation, install short lengths of tile at an angle instead of making a right angle in the tile or pipes.

Going Up

To build the foundation walls, rely on a foundation company that uses plywood forms reinforced with reusable 2 x 4s. Forms must be braced, and must be both plumb and level. The level requirement goes two ways: across the width and along the length of the wall. The forms must be built to accommodate pockets for steel or wood girders, which will hold up floor joists in the middle of the addition. Forms must also be built for any windows, fireplaces, doors, hatchways, and just plain jogs in the foundation.

The form men will make sure the foundation walls stop at the right level. Most of the time this will be below the tops of the forms; the concrete generally will find its own level and smoothing off is unnecessary. A proper job will make the top of the foundation fairly smooth, enough so that the wood sill will set snugly on top of it.

Concrete must be poured in one continuous operation. If you pour half the foundation walls and quit for the day, and then continue, you will have a horizontal seam in the foundation that could leak, and, if it is above grade, be unsightly.

At the top of the foundation wall, insert anchor bolts in the fresh concrete (Figure 44). Anchor bolts are ½-inch steel bolts curved at one end and threaded on the other to take a washer and nut. Embed the bolts, curved end down, into the concrete, at 4- to 8-foot intervals, with enough threaded end exposed so that it will go through the sill to allow the nut to snug the sill onto the foundation. Sill sealer, a strip of fiberglass, goes on top of the foundation walls to fill any irregularities between foundation and sill.

A termite shield is a good thing to install here. It is a metal strip put on top of the foundation to form a sort of hood over it on the outside, preventing termites from building their mud tunnels up the foundation and into the wood. Sometimes they work.

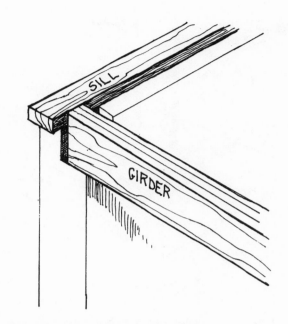

FIGURE 45. *Basement floor beam or girder sits in a pocket formed in the foundation.*

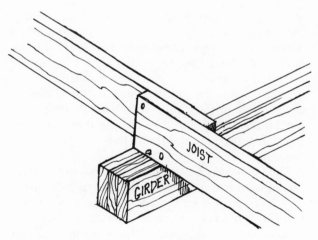

FIGURE 46. *Floor joists overlap on the girder and are nailed.*

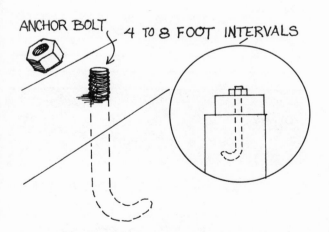

FIGURE 44. *Anchor bolts are inserted in the top of the foundation. They hold the wood sill in place (inset).*

If your addition is wider than 15 feet, it is a good idea to install a steel or wood girder in the center, going from one wall of the foundation to the opposite wall. The girder is set into pockets (Figure 45) so that its top is flush with the top of the sill. Floor joists will then span half the foundation's width, overlapping on the center girder (Figure 46).

From the Bottom

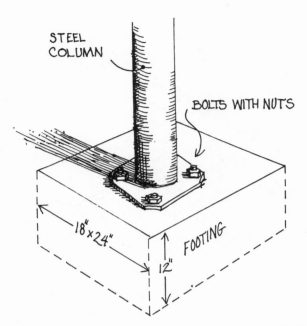

FIGURE 47. *A column or post holding up the basement beam must have its own footing.*

Steel columns must also support the girder, generally every 6 to 8 feet. Each column or post must have its own footing (Figure 47), poured at the same time the footings or foundation is poured. Column footings are usually 12 inches deep and 18 to 24 inches square. If columns were set on the concrete floor, they would be under pressure from the addition above and might be driven right through the thin floor. If they were put on the bare ground under the floor slab, and the slab poured around them, they would sink into the earth.

Steel columns have top and bottom plates to be bolted onto the girder and the concrete footings. The bolts are set in the concrete as it is poured. (The bolts are not fully necessary since, if the tops of the footings are below the final level of the concrete floor, the concrete will surround the bottoms of the columns and anchor them.) Wood posts also can be used, but should be made of pressure-treated wood. A steel pin is inserted in the concrete as it is poured and the post, which has a hole drilled in its bottom end, is inserted over the pin. Sometimes a pedestal of concrete is used, particularly if no floor is poured.

Once the posts are in place, the concrete basement floor can be poured (by a professional, please!). Although the actual pouring may be one of the last things to do on the project, short of finish work, now is the time to consider the floor because of insulation needs.

First, put down 6 to 18 inches of crushed stone. This will help prevent water from rising against the slab. When that happens, the pressure is irresistible.

Then install 6-mil polyethylene on top of the crushed stone (Figure 48). This is a vapor barrier that will keep ground moisture from penetrating the slab. Install 4 inches of Styrofoam insulation on top of the vapor barrier, giving an R factor of about 20.

The slab itself should have 6-inch steel reinforcing mesh embedded midway in its 3-inch thickness. Put down stones to hold the mesh in place as the slab is poured. The floor should be smooth, troweled with a steel trowel.

The floor should slope toward any drains, such as storm-sewer drains. And to keep ground water

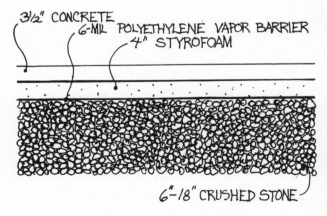

FIGURE 48. *Treatment of basement slab or slab on grade: crushed stone for drainage, 6-mil polyethylene as a vapor barrier, 4-inch Styrofoam insulation, and 3½ inches of concrete.*

Excavating and Foundations

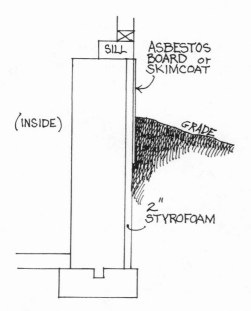

FIGURE 49. *Glue Styrofoam to the outside of the foundation, and put a protective coating over it above grade.*

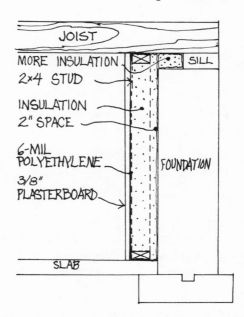

FIGURE 50. *Inside insulation on a basement (foundation) wall.*

away from the basement, particularly where ground water is high, a sump should be installed. A sump is simply a hole in the floor 18 inches square and 18 inches or so deep, with concrete sides and a gravel bottom. The floor of the basement should slope toward the sump. Then if flooding occurs, the water will go into the sump, from which it can be pumped by an automatic sump pump. Water can be pumped through plastic pipe to at least 10 feet away from the house or into the storm sewer or dry well, which also should be at least 10 feet away.

The sump also acts as a relief valve for any ground water, which can fill the sump and be pumped away before it floods the floor or exerts concrete-breaking pressure on the underside of the slab.

And, speaking of water, the foundation walls are waterproofed on the outside before being insulated by applying a layer of hot tar from the footing to grade level. Then apply a layer of 15-pound roofing felt, another layer of tar, a second layer of felt, and a third layer of tar; overlap the felt by half its width.

Now apply two inches of Styrofoam on the outside of the foundation, over the waterproofing material (Figure 49). Use a construction adhesive in a caulking cartridge or an adhesive specifically designed for gluing Styrofoam. The Styrofoam goes from the footings to the sill. Because it is flammable, it must be covered with fire-resistant material above grade. Below grade, the backfill is enough. Use asbestos board or a fire-resistant skimcoat of concrete above grade.

If the basement is not going to be occupied or lived in, the outside insulation is sufficient to prevent the cold from penetrating the concrete. Also, with outside insulation, the warmth from the basement will not only be retained in the basement itself, but also will be absorbed, to an extent, by the concrete.

It is not necessary to insulate the inside of the basement walls, particularly if the basement ceiling is to be insulated. But if desired, and definitely if the basement is living space, build a stud wall on the inside of the foundation to hold insulation.

Build the 2 x 4 stud wall, with the studs on 24-inch centers, about 2 inches from the foundation (Figure 50), and insert 6-inch batts or rolls of unbacked fiberglass insulation in the wall cavity, so that the fiberglass is touching the concrete. Apply 6-mil polyethylene on the face of the wall toward the basement and cover with ⅜-inch plasterboard. The stud wall can be built simply, with a single floorplate and a single top plate. The plasterboard can be taped and compounded, or skimcoated, or, if the basement is unoccupied, left as is.

From the Bottom

Remember, for a full basement foundation as described here, it is best that a professional crew build the forms and pour and finish the concrete for the foundation walls. If you really want to do it yourself, put down concrete blocks.

Concrete Block Foundation

Laying concrete blocks is not easy. The blocks weigh 30 pounds each. They are nominally 16 inches long, 8 inches high, and 8 inches wide. (Some codes require 12-inch-wide blocks, with the other dimensions the same, for foundations.) Subtract ⅜ inch from the dimensions of a block to get the true dimensions. A ⅜-inch mortar joint (horizontally and vertically) will bring the block up to full dimension.

The best block formation is a common bond, with the second course (layer) offset by half the length of the block in the course below.

Blocks (and bricks, for that matter) are laid up in cement mortar, made of cement, hydrated lime, and sand. Don't use beach sand; it has salt in it, an enemy of concrete. In fact, specify mortar sand when mixing mortar; it is finer than ordinary sand used in concrete. For regular service mix 1 part portland cement, 1 to 1½ parts lime, and 4 to 6 parts sand. Better yet, use 1 part masonry cement or mortar cement (it already has the lime in it) and 3 to 4 parts sand. You can buy ready-mixed mortar, called mortar mix, and add only water, but it is expensive. Mortar should be on the dry side; too much water will make it soupy, sloppy, and weak. Dryish mortar tends to keep off the face of the block. It should be pliable enough to adhere to the blocks without falling off. It should be firm enough so that when a block is put down on a bed of mortar, its weight will not squash it or make it ooze too much out of the joint.

When you use concrete blocks, you do not need a key in the footing. It would be a good idea to

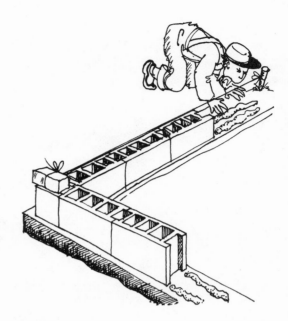

FIGURE 51. *Starter course of concrete blocks on a concrete footing.*

stick a few short pieces of reinforcing rod into the footing, at 4- to 8-inch intervals, which will stick up into the holes of the blocks. When the block is laid, the hole with the steel rod is filled with concrete or mortar.

Don't mix more mortar than you can work with in about half an hour. It will begin to set up in 15 minutes, and if it stands for much longer than half an hour, it will be unworkable. If it gets a little too thick, add some water and remix. It will be weakened slightly, but will be all right to use.

To start, lay a bed of mortar (two rows corresponding to the front and back of each block), or a full bed, on the footing (Figure 51). Start laying blocks at a corner. Build several blocks out from the corner, and several courses high. Butter one end of a block and set it into place. When a second block is set, its buttered end is placed against the end of the block before it (Figure 52). As each course is laid, butter the top of the course below it. Use more mortar than you need; it is easier to hunker down a block into the right position than to add extra mortar to build it up to the right position. Build a second corner after the first course has been laid along one side of the foundation; build a third and a fourth, or as many as the foundation makes up. The corners will act as a guide for the wall built up between them.

Excavating and Foundations

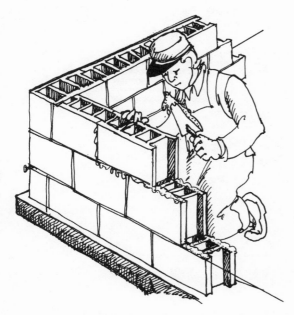

FIGURE 52. *Corner treatment with concrete blocks: the nail holds a string as a guide for the height and alignment of one course of blocks.*

Use a 4-foot mason's level to make sure everything is level and plumb. An 8-foot one is better, but you can use a straight 2 x 4 as a straightedge. Run a string (Figure 51) from the top of the first course at one corner to its counterposition at the other corner. This is only a guide, because a person working blocks can easily push the string out of position without realizing it. Every few blocks, check for level and plumb, then check what you have finished—both the top of each course and the inner and outer vertical face—with the long straightedge to make sure the wall does not undulate. If the straightedge touches the wall or top of the course for its full length, you've done a yeoman job. If there are humps or gaps revealed when the straightedge doesn't touch the wall all along its length, you are wavering, and while not much can be done short of taking the wall apart and starting over, you can be more careful as you continue.

The tops of the wall must be capped by a solid cap 4 inches thick or by concrete reinforced with wire mesh. You can also finish off the top by filling the holes with concrete or mortar. Insert anchor bolts every 4 to 8 feet by filling the top two blocks with concrete or mortar and inserting the bolt in the right position. To keep from filling the whole wall with concrete, put a piece of screen or hardware cloth at the bottom of the second from the top course, or wedge in a stone.

A concrete block wall must be protected by a coating of mortar, then waterproofed as described above for concrete walls. Insulation is applied in the same way as with the concrete walls, too. In fact, if you plan to put Styrofoam on the outside of the walls, it can be embedded into the final coating of tar, below grade. Not only will the tar hold it, but the backfill will apply continuous pressure on it as it fills the trench.

Having built your basement, whether concrete or concrete block, and capped the foundation walls with sill sealer, termite shield, and sill, you may want to connect this basement with an existing one in your house, by cutting a door opening. The foundation of the house should be cut by a professional cutter. All you need is a regular door opening; the sill on top of the foundation is sufficient to span the opening.

CRAWL SPACE WITH FOUNDATION

A crawl space is probably the best bet for an addition. It needs a foundation, but one that needs to go only below the frost line (3 to 4 feet and more in cold climates, less in warmer ones). There is much less excavating necessary as compared to a full basement, although you should remove enough earth from the crawl-space floor to allow 3 feet between the floor and the bottom of the floor joists above.

A crawl-space foundation can be as little as 6 inches thick, but 8 inches is better, with footings 8 to 12 inches deep and twice as wide as the walls. Concrete blocks can be 8- or 12-inch.

Instead of steel or wood columns to hold up interior beams, short columns on footings or short piers of concrete are built. Sonotubes, cardboard

From the Bottom

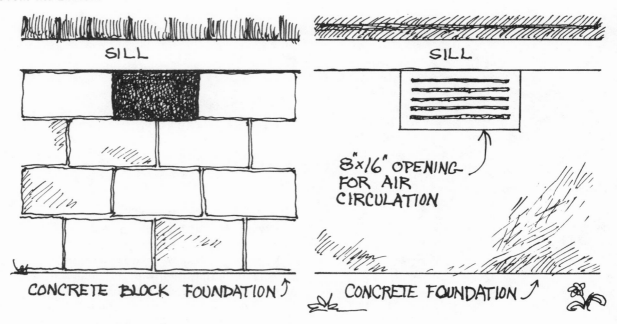

FIGURE 53. *To ventilate a crawl space, 8-by-16-inch openings at the top of the foundation will permit adequate ventilation. In a concrete-block foundation, the opening is created by omitting one block. In a poured concrete foundation, the opening is created as the concrete is poured.*

tubes that come in various diameters, fill this need nicely (see Figure 55). After the footings are installed, the Sonotube is set in place and concrete poured into it to the correct level. The cardboard is torn off after the concrete sets. Concrete-block piers can also be built, by laying one concrete block on top of another. The piers must stop at the right level, and their tops must be level with one another. If they are not, install shims to snug the piers and beam together.

Most basements are ventilated in one way or another, usually by openable screened windows that can be left open in summer. A crawl space is ventilated, or should be, so pockets are installed in the forms for the concrete wall to accommodate vents. With concrete blocks, a block can be left out every 8 feet or so on top of the wall, and the hole covered with screened, louvered ventilators. An alternative method is to put a block on its side in the opening, with its holes facing out, and screen it (Figure 53). The ventilators are left open all year long; some experts recommend closing them in the dead of winter. Ventilation is required to prevent the buildup of moisture in the crawl space, where it could condense on cold members, causing mildew and possible decay.

The floor of the crawl space is not paved, but the earth is smoothed out and covered with 6-mil polyethylene overlapped 2 to 3 feet. Roll roofing can also be used, overlapped 6 inches and sealed with roofing cement. The polyethylene or roll roofing will prevent moisture from coming up from the earth, which can happen whether the earth in the crawl space feels dry or not.

With the soil cover, the ventilation requirement is not great: 1 square foot of ventilation area for each 1,600 square feet of crawl-space floor area. If no soil cover is used, however, this ratio plummets to 1 square foot of ventilation area to 160 square feet of floor area, because of the moisture from the earth.

A crawl-space foundation is insulated on the outside in the same way as a basement foundation. There is no need for insulation on the inside of the walls. Insulation also is installed on the crawl-space ceiling.

Crawl Space with Piers

The simplest and least expensive foundation is made of piers of concrete, with large beams or girders set on them. The piers take the place of concrete or concrete-block foundations.

The number of piers depends on the size of the addition. Piers that support girders should be

Excavating and Foundations

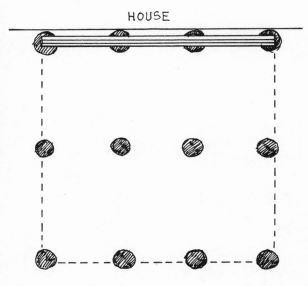

FIGURE 54. *When piers are used as a foundation, they are set at least 7 feet apart to carry the girders that will support floor joists. The girders are 10 feet apart. The one next to the house is not necessary if a ledger is applied to the house sill or wall as a support for the joists.*

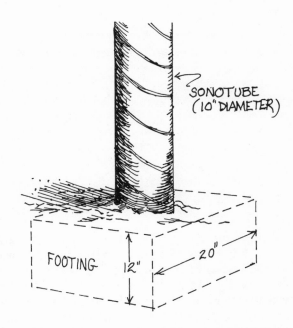

FIGURE 55. *A Sonotube holds concrete, creating a pier.*

placed every 5 to 8 feet, depending, again, on the size of the addition. If it is 20 feet wide and 20 feet long, the piers should be placed every 7 feet (Figure 54). There would be four piers 7 feet apart along the 20-foot width of the addition where it butts up against the existing house, four more placed halfway down the 20-foot length of the addition, and four along the outer edge. A girder would sit on each set of four piers: three girders, therefore, for a 20-by-20-foot area. Three 2 x 12s spiked together make a good girder.

Sonotubes make the best piers. Dig holes below the frost line and build concrete footings at the bottom of each hole: they should be square to a measure twice the diameter of the Sonotubes and 12 inches deep. A 10-inch Sonotube is adequate for the pier. Cut the Sonotube to length, set it on the footing, and backfill a little to hold it in place (Figure 55).

When it comes to determining the height of the piers, you have to measure carefully. If the floor of the addition is to be the same as that of the house, measure from the house floor to the top of the footings. Subtract the thickness of the floor, subfloor, joists, and girders, and you have the height of the piers (Figure 56).

Once the Sonotubes are in place and plumb, pour the concrete. After a day, backfill and tear away the cardboard of the tubes above grade. The rest will rot harmlessly away. You can put an anchor bolt in the top of each Sonotube, or embed a steel fastener in each pier that will be nailed to the girder (Figure 57).

The problem with a pier foundation is that the space under the floor is open, and may attract vermin or other pests. The sides of the open space can be covered with concrete, asbestos board, or aluminum skirting material. Possibly better, and more attractive, is 1 x 6 or 1 x 8 tongued and grooved pressure-treated wood boards set verti-

From the Bottom

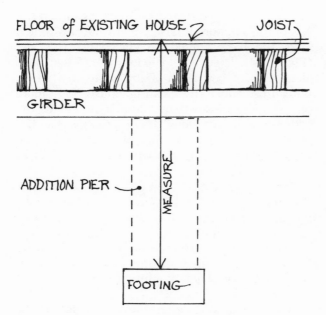

FIGURE 56. *Measure carefully to determine the height of a pier or concrete column holding up a basement girder.*

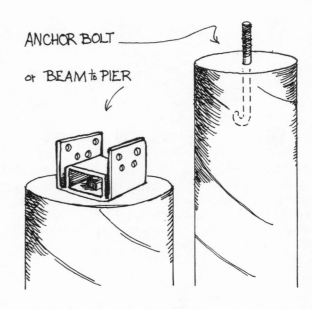

FIGURE 57. *Two ways to connect a pier to a beam.*

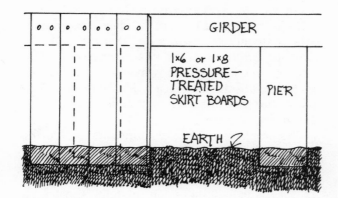

FIGURE 58. *Covering a crawl space with skirt boards.*

cally into the ground and nailed to the girder and joists (Figure 58). This sounds like a terrible way to use wood, but being pressure-treated it will resist decay and is also insect-resistant. Another way to cover the sides is with wood lattice. As you can see, despite the ease of buiding and the small cost, the pier foundation is probably the least desirable.

Slab on Grade

Another foundation is a slab of concrete on grade: that is, one built directly on the ground with a gravel underlay.

A slab on grade is not an ideal floor because it is hard, but it *is* an ideal heat sink, to absorb and store heat from the sun when it's shining in, and to radiate heat into the room when the sun isn't shining. A 4-inch layer of Styrofoam insulation under the slab prevents the gravel and earth from absorbing the sun's heat. If that happened, the concrete would never get warm. In other words, the heat sink would be too large, destroying its ability to store the right amount of heat and to radiate it when it is needed.

There are two ways to build a slab on grade, both with foundations that go below the frost line to prevent the slab from moving in freezing and thawing weather. One type is a monolithic (literally, one stone) slab, which combines footing, foundation, and slab all in one piece, or pour, of concrete. The other is made with separate footings, foundation, and slab. Both are good.

The slab should be high enough above grade to allow for the ground to slope away from the foundation, and to prevent the wood of the walls from being in contact with the earth. Eight inches is minimum, 12 inches better. The slab should be 4 inches thick, laid over 6 to 12 inches of gravel or crushed stone, and reinforced with steel bars or mesh.

For a monolithic slab, excavate a perimeter ditch about 12 inches wide and deep enough so that the bottom of the ditch is below the frost line. The 12-inch width is a minimum one for digging. It's useful to get a number of small but strong children to dig a trench of this type, because they can fit best in a 12-inch-wide trench. Footings can be used with a monolith, but they are not neces-

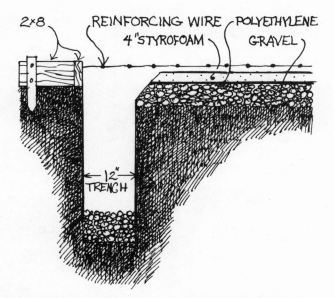

FIGURE 59. Preparation for a monolithic foundation and slab, poured all at once in one piece.

sary. The foundation starts at the bottom of the ditch, on undisturbed earth. If the earth is not too sandy, the walls of the ditch will be the form that holds the concrete. Above grade, set a 2 x 8 or 2 x 12 on the ground on the outside of the ditch, securing it with stakes (Figure 59). Since you cannot drive stakes next to the board, because it is right on the edge of the ditch, build 2-foot extenders and attach one end to the form board and the other to a stake driven about 2 feet away from the board.

From the Bottom

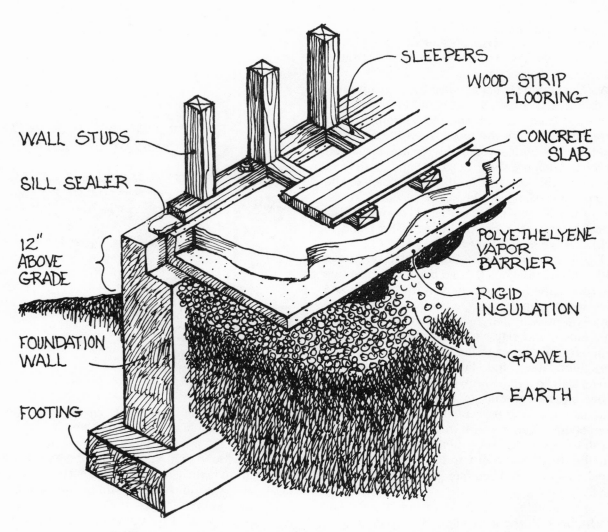

FIGURE 60. *A slab on grade poured separately from the foundation.*

Excavating and Foundations

On the inside of the ditch, excavate deeply enough to accommodate 6 to 12 inches of crushed stone, a 6-mil polyethylene vapor barrier, 4-inch Styrofoam insulation, a 6-inch steel mesh, and the 4-inch slab. The mesh must go in the middle of the 4-inch slab thickness. Two-inch Styrofoam is applied to the outside of the foundation from the top of the slab to several inches below grade, and covered above grade with a fire-resistant cement-based coating.

Now pour concrete into the ditch and onto the mesh, which has been set on stones or fresh concrete to bring it midway in the thickness of the slab. With a partner, screed the slab smooth by pulling a 2 x 8 or other large, long, and heavy board across the concrete, back and forth, using the edges of the forms as a resting platform and a guide.

Install anchor bolts every 4 to 8 feet in the slab, to allow for snugging down of the wood sill.

For a slab on a separate foundation wall, the wall is set on footings and is notched at the top to accommodate the slab, which is poured separately (Figure 60). The footings and foundation are poured much like those for a crawl space, complete with anchor bolts. The foundation walls are insulated on the outside. Then dig out inside the perimeter, going deep enough to accommodate the 6 to 12 inches of crushed stone, 6-mil polyethylene, 4-inch Styrofoam, and the 4-inch slab.

The slab can be smoothed off in one of two ways: if a wood floor is planned on top of the concrete, it is troweled with a wood float, beginning 15 minutes to half an hour after the concrete is poured. If a tile floor is planned, then the concrete should be troweled with a steel trowel, which makes it very smooth.

To make a slab a heat sink for solar heat, any finishing floor material must be masonry, such as glazed or unglazed clay tile or bricks. A wood floor would act as insulation, preventing the solar heat from warming the concrete and the concrete from giving up the heat later on. The same is true for wall-to-wall carpeting; if there are going to be rugs, they should be area carpets.

Any of the procedures described in this chapter for building the various types of foundations is applicable whether the addition is on the same level as the existing house or on a different one. The addition can be a step or two up or down; more than that, however, and there might be complications because of the need for a stairway and consequent problems with existing ceiling heights.

chapter 5

It's a Frame-up

Floors

With the foundation in place, it's time for the builder—you—to go to work, because from now on most of the work is with wood, and the amateur can do well with this kind of construction. In dealing with beams, sills, girders, and joists, though, get a helper, who can cut the physical labor in half or more. Just try wrestling with a 20-foot 4 x 6 wood sill by yourself!

Figure 61 shows wood construction, in this case platform construction, where each floor is a platform on which the walls for each succeeding floor (story) are set. Joists, studs, and rafters are set on 24-inch centers (denoted as "24 inches o.c.," for "on center"). Check to see if this agrees with your community building code. Standard construction calls for members to be set 16 inches o.c., and while the 24-inch system is as strong as the 16-inch, some codes may not allow it. The 24-inch system can save several hundred dollars in a standard-built house. And since this addition is superinsulated, with double walls, the 24-inch system is important to make up for the cost of the extra material in the double walls.

The first wood that will be put on the foundation is the sill or sill plate (Figure 62), a 2 x 6 or two 2 x 6s spiked together. (Some codes may require a 4 x 6, in one piece, shiplapped at corners.) But wait—some nonwood items must be put on before the sill.

First install the sill sealer, a thin piece of fiberglass that seals any irregularities between foundation and wall. Next apply a termite shield, which is a piece of 12-inch aluminum or copper flashing set on the foundation and overlapping it on both the inside and the outside by one inch, if the foundation is 10 inches thick, and by 2 inches if it is 8 inches thick (see Chapter 4). Now you put on the sill, ¾ inch in from the outside of the foundation, to allow the sheathing to butt down against it. If the sheathing is ½-inch plywood, make the space ½ inch.

When rigid insulation is put on the outside of the foundation, the approach is different (see Figure 18). Suppose 2 inches of rigid Styrofoam is glued to the outside of the foundation. Then set the sill so that it overlaps the foundation by 1¼ inches, and covers half the insulation. The sheathing and the siding will bring the outer skin out far enough to overlap the insulation and any protective material that covers it. Flashing can be inserted behind the sheathing to protect the insulation. This flashing serves the same purpose as the termite shield and can replace it.

Drill holes in the sill to coincide with the anchor bolts in the foundation (see Figure 44). The sill is slipped over these bolts and secured with washers and nuts. Don't snug them down too tightly; the sill should be straight and level. The sill is

FIGURE 61. *The anatomy of a double-walled, wood-framed building.*

KEY:
1. Foundation
2. Anchor bolt
3. Sill sealer (fiberglass)
4. Basement column or pier, on footing
5. Center basement beam or girder
6. Sill
7. Floor joist
8. Header for opening in floor
9. Cross-bridging
10. Solid bridging
11. Floor sheathing
12. Floor plate
13. Wall stud
14. Corner brace
15. Jack studs (doubled or tripled at window and door sides)
16. Top plate (doubled)
17. Headers for windows and doors
18. Wall sheathing
19. Ceiling joists
20. Rafters
21. Ridgeboard
22. Collar tie
23. Gable studs
24. Roof sheathing
25. Roof shingles
26. Siding shingles
27. Wall underlayment (roofing felt)

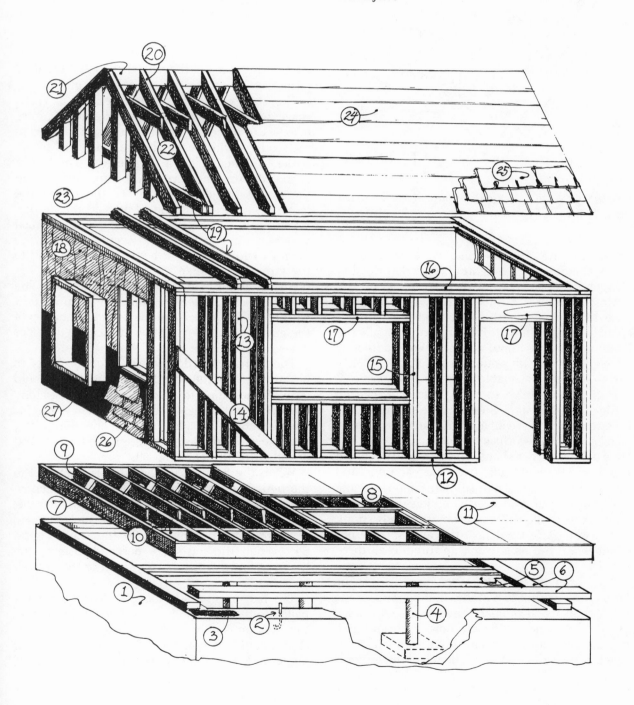

43

It's a Frame-up

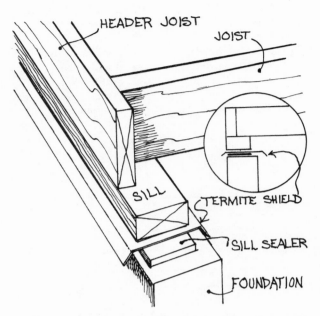

FIGURE 62. *Sill sealer and termite shield go under the sill on top of the foundation.*

attached in the same way on crawl-space foundations. The floor on a slab on grade is described in Chapter 16.

Most structures are too wide or deep for floor joists to span one length. So a large beam or girder splits this space in half. Fifteen feet is usually the maximum span without a girder. If an addition is 25 feet wide, a beam would cut this in half, to 12½ feet — a more reasonable span for floor joists.

A girder, as we've seen in Chapter 4, can be made of wood or steel. Steel is more expensive but is better. If it is of wood, it should be made of three pieces the same width as the joists: if 2 x 10 joists are used, the girder should be three 2 x 10s spiked together. Use 20d nails, two near each end and the others staggered on 16-inch centers. If the nails go through two boards, clinch them (bend their points over); this makes an unusually strong connection. The third board is also spiked, but chances are the nails won't go through all three boards.

The girder sits in pockets (see Figure 45) in the foundation, with 4 to 6 inches of it bearing on the foundation. Treat the part where it sits with wood preservative, and to prevent moisture from building up and causing decay leave ½ inch of air space at the sides and end. The top of the girder is even with the top of the sill plate, so that the floor joists can be set on the sill and the girder.

The steel columns in the basement or crawl space, which support the girder (Figure 63), have been set in place on their footings and secured to the footings with bolts. They are attached to the girder by plates held with lag screws. The height of the columns is determined by measuring from the top of the sill vertically to the top of the footings, subtracting the depth of the girder. Wood posts can be used, but only with a wood girder; use 6 x 6 posts and secure them to the footings with metal post bottom fasteners and to the girder with angle irons and lag screws.

A steel girder is installed in the same way. The top plates of the steel columns are secured to it with bolts. A 2 x 6 or 2 x 8 is bolted on the top of the girder to act as a nailing surface for the floor joists.

The joists of the addition will, of course, need to be attached to the existing house. You should strip off the siding to expose the sheathing along the area of the house where the addition is to go. It wouldn't hurt to take off the sheathing, too, because it will be easier to determine the size and condition of the existing sill.

When the joists of the addition run parallel to the house, the first one is simply butted against the house joints or the sheathing (if it is left on) and nailed; or set a fraction of an inch away to accommodate irregularities in the wall (Figure 64). When the joists are perpendicular to the house wall, they must have a girder or ledger to sit on.

The ledger is easy to install. It can be a 2 x 4 nailed to the house sill (Figure 65) if the floor of the addition is to be at the same level as the floor of the house. If not, the ledger must be located higher or lower. If it is impossible to use a ledger,

Floors

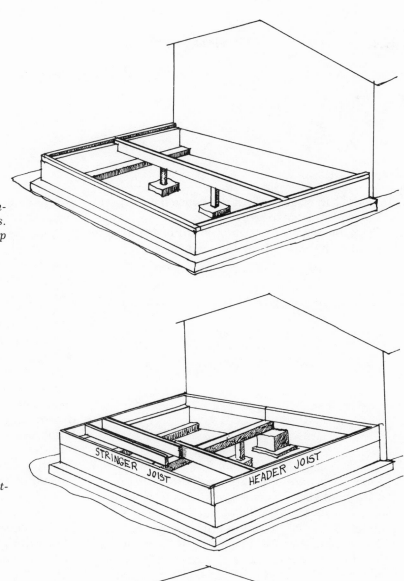

FIGURE 63. *Center girder on the foundation supported by it and by columns. The top of the girder lines up with the top of the sill.*

FIGURE 64. *Joists parallel to the existing house.*

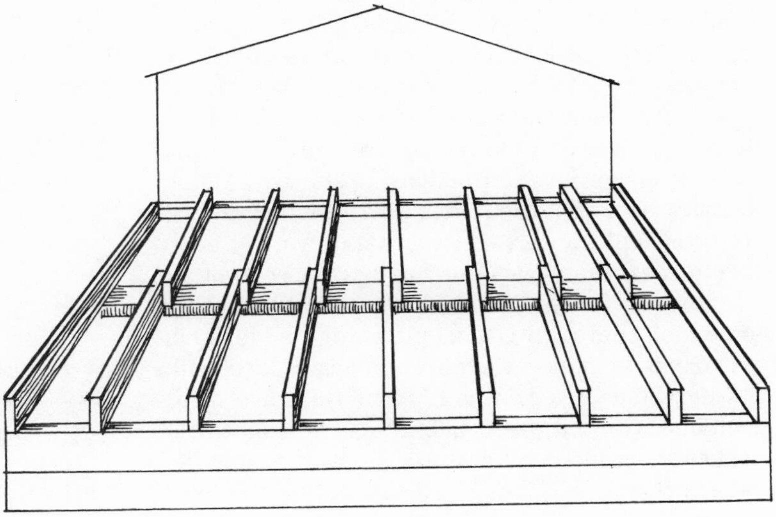

FIGURE 65. *Joists at right angles to the existing house, their ends sitting on a ledger nailed or applied to the house with lag bolts.*

It's a Frame-up

the joists can be set against the house header joist or sill with joist hangers.

The table shows the size of joist needed for its span and the space between each joist.

Joist size	Spacing (inches o.c.)	Maximum span (feet)
2 x 6	16	8
	24	7
2 x 8	16	12
	24	11
2 x 10	16	14
	24	12
2 x 12	16	16
	24	14

Floor joists may vary a little in width, although a 2 x 10 should be 9½ inches wide. But to avoid any discrepancies, they should be put in place without nailing. Start by setting perimeter joists: the stringer joists and the header joists (see Figure 64). The stringer joists will parallel the intermediate joists, and the headers will be at right angles to them. Header joists are secured by nailing through them into the ends of the stringer joists (and later, the intermediate ones). With the perimeter joists in place, you have an open box.

Now place the intermediate joists, on 24-inch centers (you can place them on 16-inch centers if you wish). If they are so long that they must be in two pieces, the pieces butt up against each other (Figure 64) and overlap by the width of the girder on which they rest. Use a long straightedge, such as a 2 x 8, to make sure that each joist is level with its neighbor. If a joist is high, it will have to be notched where it meets sill, girder, or ledger. If it is low, it is lifted by a shim, such as a shingle, which is tapered. One shim can adjust the level from ¹⁄₁₆ to ¼ inch; two can adjust to ½ inch. A discrepancy in joist level of ⅛ inch is allowed.

Place the joists with their crown edges up. You can tell which edge this is by sighting down the joist; you'll notice a slight bend in the board, and you'll see that the grain forms a raised curve — the joist crown. Put this edge up; the crown will level off with the weight of the floor and the rest of the addition above.

Now you can nail the joists. Nail through the overlap, using 12d or 16d nails, and clinch the nails. Toenail (nail at an angle) the joists where they sit on the sill and the girder, nail through the header into the ends of the joists, and toenail to the existing house joist or ledger (see Figure 65). If you use joist hangers, nail through the hanger into the existing house and through the hanger into the joist of the addition. Two methods of toenailing are shown in Figure 66.

Mark the sills and beams where the joists will go. You can space the marks properly by measuring 24 inches from one side of the stringer joist. Then put the same side of the second joist on this mark. Another way to do it is to cut a spacer block 22½ inches long: this is the distance between the joists. The spacer block will also help hold the joist in position while you toenail it (Figure 67).

As you nail the joists into place, and you come

FIGURE 66. *Toenailing techniques.*

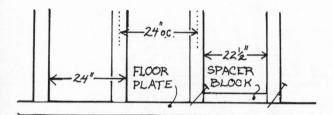

FIGURE 67. *Spacer block not only allows proper interval between studs, but also acts as a stop to allow accurate toenailing.*

to the side opposite from where you started, you must place the last joist less than 24 inches from the stringer joist. So even if you have only 26 inches to go, you must place a final joist so that it is 24 inches o.c. from its neighbor, even though it will be only an inch or two from the stringer joist.

Where openings in the floor are required, for stairways, chimneys, fireplace, and so on, the joists are doubled (Figure 68). Cross-joists called headers are placed where the joists stop and are doubled. Stringer joists, parallel to the opening, are also doubled. First the single stringers and headers are nailed, then the doubled joists are nailed to the members already in place. Joists are also doubled under partition walls.

Bridging is recommended between joists midway in their span (Figure 69). Cross-bridgings are 1 x 3s cut at an angle and toenailed. Solid bridging is comprised of 2-inch lumber the same width as the joists. To allow direct (face) nailing, bridges can be staggered. Some experts feel bridging doesn't do any good, because the wood subfloor and top floor hold the floor system together. We feel that bridging prevents joists from warping and a tendency to lean. This joist and sill installation permits a full subfloor on which the two parts of the doubled superinsulated walls will sit (see Figure 22).

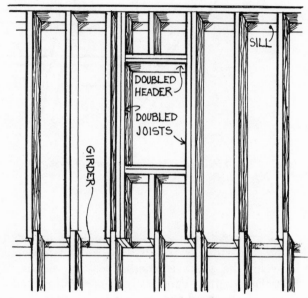

FIGURE 68. *Joists are doubled for openings; so are header joists.*

It's a Frame-up

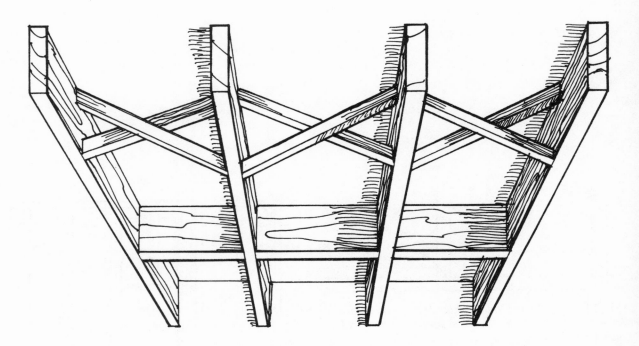

FIGURE 69. *Two kinds of bridging: cross-bridging (foreground) and solid bridging.*

Superinsulation in the floor should be installed before the subfloor is put down, while the area between the joists is accessible. Six inches of fiberglass, with a vapor barrier of 6-mil polyethylene on top, should be secured with chicken wire stapled to the bottom of the joists or with "lightning rods" (see Figure 19). If the floor joists are more than 6 inches deep, there will be an air space above the insulation and vapor barrier, which serves to increase the insulating value of the floor. A basement ceiling should be installed, of plasterboard or ceiling tiles, to prevent agitation of the fiberglass and its possible infiltration into the basement area. This is especially important if the basement is to be used for living space. If not, the ceiling doesn't have to be finished off.

Now comes the subflooring. It can be boards or plywood, both of which are good, whether the finish floor is hardwood boards or planks or a plywood underlayment for carpeting, resilient tile, or ceramic tiles. Boards are nominally 6 to 8 inches wide, 1 inch thick, tongued and grooved, and laid diagonally, at a 45-degree angle to the joists. Any other way would make it difficult to lay finish-floor boards on top of them (see Chapter 16). Plywood should be ⅝-inch thick, with exterior (water-resistant) glue; plyscore is a good grade to use. Boards are nailed with two 8d or 10d nails. Butt board ends over a joist. Lay them so that they overlap the perimeter of the addition, then cut them with a rotary saw, flush with the stringer and header joists.

For plywood, install 4-by-8-foot sheets with the grain at right angles to the joists. Stagger joints. Nail with 8d hot-dipped zinc galvanized nails or ring-shanked nails. Don't use screw nails, which may tend to work loose. Space the nails 6 inches apart along the edges and 10 inches apart on the intermediate joists under the inner areas of the plywood sheets. Space joints 1/16 to 1/8 inch apart to allow for expansion.

Hardwood boards are nailed on top of the subfloor. Tiles and carpeting are laid on ½-inch underlayment-grade plywood. They should not be laid directly on the one-layer subfloor; it is not strong enough and will tend to flex under weights and stress. If ceramic tile will be the final flooring, the joists should be set on 12-inch centers and the plywood subfloor should be ¾-inch thick.

chapter 6

The Uppity Parts

Walls

When building an addition, questions come up: when should you remove the siding and sheathing of the existing house, where the addition will go, and how much should you remove? As we have seen in the preceding chapter, the attaching of the floor joists necessitated removal of the siding at the joist area, and probably some of the sheathing. The matter becomes more important in connection with walls (and later with the roof).

In general, stripping of siding and sheathing depends upon the design of the addition and the progress of the work. You want to protect your wall from the weather, so the siding and sheathing should be left on as long as is practical. You may need to remove only the siding and leave the sheathing intact to allow walls and other parts of the addition to be nailed to it. The sheathing can also be used as a base for plasterboard and a finished wall surface. If, however, it is uneven, it can be removed and new plasterboard put on. Opening the house wall, too, will ease the installation of any electrical wiring and plumbing. As we continue to build the addition, specific details about the attachment to the existing house will be covered.

For a superinsulated addition, the walls are especially important. Here we will arbitrarily describe one way to build superinsulated walls, although there are any number of variations. No matter how the wall is built, however, it is of a certain minimum thickness and contains enough insulation for an R factor of 30 or greater. Our wall is doubled. Each wall is made of 2 x 4 studs 24 inches o.c., with the studs of the outer wall offset by 12 inches from the studs of the inner wall (Figure 70). The walls are 4 inches apart, so that the entire wall assembly is 11 inches thick, from the edge of one wall to the edge of the other. A 6-inch batt or roll of fiberglass is then installed in each wall, one layer on the outer wall and one on the inner, with the inside wall covered with a 6-mil polyethylene vapor barrier. The slight squeezing of the insulation in this "sandwich" will not reduce its R 38 factor by much. Half-inch plywood is applied to the outer wall; ½-inch plasterboard is put on the inside one and given a ⅛-inch skim-coat of plaster—bringing the total thickness of the doubled wall, not counting exterior siding, to 12⅛ inches, and the R factor to nearly 40.

This is admittedly very thick, but with today's insulation, it is the best way to achieve superinsulation. The interior wall, by the way, is not a bearing wall and so there is no need to reinforce the floor below (Figure 71), particularly when the wall is perpendicular to the joists. Where the wall is parallel to the joists, an extra joist 9½ inches away from the stringer joist can be installed for

The Uppity Parts

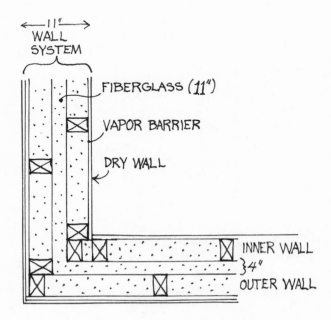

FIGURE 70. *A double wall of 2 x 4 studs set 4 inches apart, making an 11-inch-thick wall, not counting exterior and interior wall finishes.*

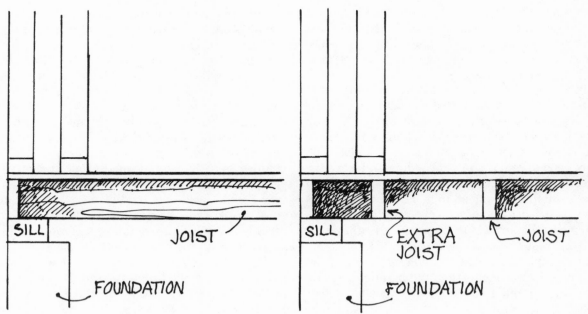

FIGURE 71. *Double wall over joists at right angles to the foundation needs no extra reinforcement (left); over joists parallel to the foundation, an extra joist is inserted to support the inner wall (right).*

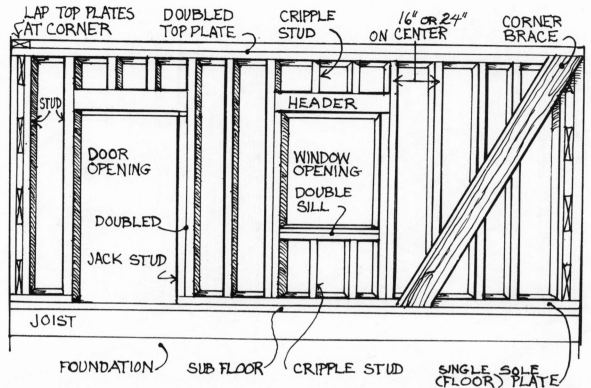

FIGURE 72. *The anatomy of a stud wall.*

extra support. Install insulation between this joist and the stringer before putting on the floor.

Whether a wall is double or single, on 24-inch centers or made of studs other than 2 x 4s, it is still put together in the standard way and is of standard height (Figure 72).

A wall is 8 feet high, from subfloor to the top of the doubled top plate. Because there are three 2 x 4s built into the wall—one floor plate and two top plates—that total 4½ inches, subtract this measurement from the length of the studs. Cut them at 7 feet 7½ inches. The final height of the walls from finished floor (¾ inch) to finished ceiling (1¼ inches) will be 7 feet 10 inches.

Cut the studs to the appropriate length. Lay the floor plate and one top plate flat on the floor, using blocks or an extra 2 x 4 nailed to the floor as a cleat. Use a tape rule or a precut piece of 2 x 4 (stud spacer) to mark stud positions along the floor and top plates. The plates will probably not be long enough to cover the width of the entire addition, so cut and lay the shorter floor-plate pieces so that they will meet at a floor joist, and cut the shorter top plates so that they will butt at a stud.

The stud spacer not only will allow you to space the studs 24 inches o.c. (22½ inches apart), but will help keep the studs in position as you nail them. Use two 16d nails to nail the floor plate onto each stud. Then nail the top plate onto each stud.

Corners are special. A double-walled corner is treated differently from a standard single-wall corner. The outside wall corner can simply be two walls butted together and nailed (Figure 73). The inside wall, to allow nailing of inside wall surfaces, is built in one of two ways: the standard technique (Figure 74) is to place a 4 x 6 on one end of one wall, or two 2 x 4s separated by short lengths of 2 x 4 spacers. A more modern technique that saves studs and allows more insulation is to put a regular 2 x 4 at the end of the wall, adding a second 2 x 4 with its wide edge facing in (Figure 75).

Building a wall without windows and door openings would be simple but impractical. Modern windows come set up; that is, already in their frames, needing only to be inserted into the rough openings of the wall. So you must know the sizes of the windows before making the rough openings, which must be about ¼ inch wider at each

The Uppity Parts

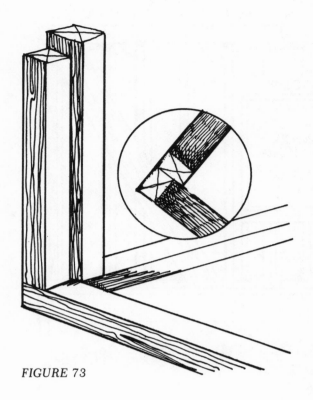

FIGURE 73

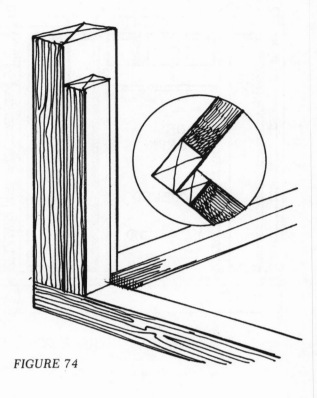

FIGURE 74

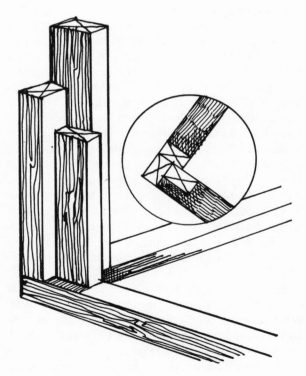

FIGURE 75

FIGURE 73. The corner of the outer wall needs no nailing surface on the inside.

FIGURE 74. Corner treatment of the inner wall uses a 4 × 6 on one wall and a 2 x 4 on the other to allow a nailing surface for the inside wall finish.

FIGURE 75. Another corner treatment, using 2 x 4s.

Walls

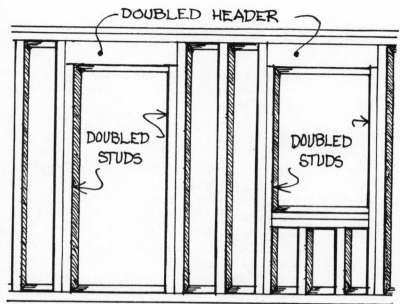

FIGURE 76. *Doubled headers over window and door.*

side and at the top to allow for leveling and plumbing the inserted windows.

Most doors are 6 feet 8 inches high (80 inches); some are 78 inches. Width will vary according to use. Setup doors and windows come with side and head (top) jambs and exterior casing. Here is how to build a rough opening for a door in both the outer and inner wall of a superinsulated wall (see Figure 72). Determine the width of the rough opening, add 3 inches, and set full-length studs at these points. Now cut two 2 x 4 studs to the height of the rough opening and nail one next to each of the 2 x 4s framing the opening. Use 16d nails. These inner studs support the header. After the wall is in position and nailed to the floor, you can cut the floor plate in the opening.

Rough openings for windows are handled similarly (Figure 72), except that they don't go clear to the floor. Doubled studs are set up; then a 2 x 4 is nailed between the supporting studs at the height of the bottom of the window's rough opening. Short studs, called cripples, are nailed on 24-inch centers to support this crosspiece, which is basically the sill of the rough opening. Toenail the sill to the supporting (side) studs with 10d nails; face-nail the sill to the cripples with 16d nails; and toenail the cripples to the floor plate with 10d nails. Rough openings can be built into the wall as it is being constructed, or built in after the wall is erected.

In most construction, the tops of windows and doors are at the same level, so headers can be installed in a standard fashion: nominal 2-inch lumber is doubled and set on top of the inner studs of the opening (Figure 72). A door needs only 2 x 6s or 2 x 8s because it is not very wide, but window headers vary according to the width of the opening:

Opening	Header size
3 to 5 feet	2 x 8
Up to 6½ feet	2 x 10
6½ feet and beyond	2 x 12

Sometimes headers are set just below the top plate, with or without cripple studs (Figure 76). This may be overly heavy construction, but the headers must support a roof or a second floor or both. In the opening for double windows usually there is a 2 x 4 stud dividing the opening and supporting the header; if, however, the double window doesn't have this divider, then the header rule above applies.

Doubling of the nominal 2-inch lumber (1½ inches thick) brings the thickness of the header to 3 inches, half an inch short of the width of the 2 x 4 studs, which are 3½ inches wide. Normally a ⅜- to ½-inch piece of plywood or plasterboard is inserted between the header boards. But in a superinsulated wall, ½ inch of High R Sheathing

53

The Uppity Parts

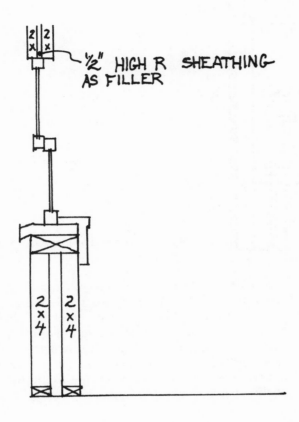

FIGURE 77. *A header of paired 2 x 8s or 2 x 10s, with ½-inch plasterboard or plywood as a filler; or, better, ½-inch rigid insulation.*

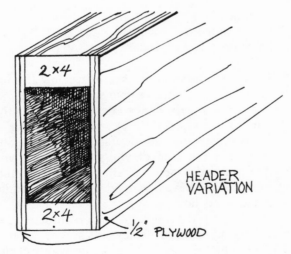

FIGURE 78. *A variation of the standard header (doubled 2-inch lumber with a filler piece): a box beam created by 2 x 4s and ½-inch plywood. Filling the void with insulation adds to the protection.*

or Thermax is substituted (Figure 77). Or instead of the doubled headers you can use a pair of box beams, each with 2 x 4 bottom and top plates connected on both sides by ½-inch plywood and filled with insulation (Figure 78).

Build the stud wall in sections so that it won't be difficult to raise into position on the floor you've already built. Raise each section on its floor plate, with the plate running along the edge of the subfloor. Nail the plate to the subfloor with 16d nails, making sure they go into the floor joists or into stringer and header joists. One nail for each joist is enough. Nail the sections together.

Plumb the wall (make sure it is vertical) at both ends and along its face, and secure with a temporary brace by nailing the brace diagonally, from near the top of a stud to a 2 x 4 cleat nailed to the subfloor (Figure 79). Braces can be put every 10 feet or so, or wherever convenient. Do all the outer walls. Where a wall of the addition butts against the existing house, nail through the end studs of the new wall into the house sheathing. Or, if you have removed the sheathing, insert horizontal cross-pieces of 2 x 4s between the existing studs to serve as nailing surfaces for the addition wall.

Now nail the second top plates of the walls to the existing ones, lapping them at the corners (see Figure 72). This helps tie the walls together. The corners are also nailed; use 16d nails.

This completes the framing of the outer walls. When they are in position, insert 6 inches of unbacked fiberglass between the studs (see Figures 22 and 70). The width of the fiberglass depends upon the distance between the studs: 23-inch insulation will fit by friction between studs set on 24-inch centers; 15-inch insulation will fit between studs on 16-inch centers. Where the studs are closer together than specified, cut the insulation about an inch wider than the gap. Once

Walls

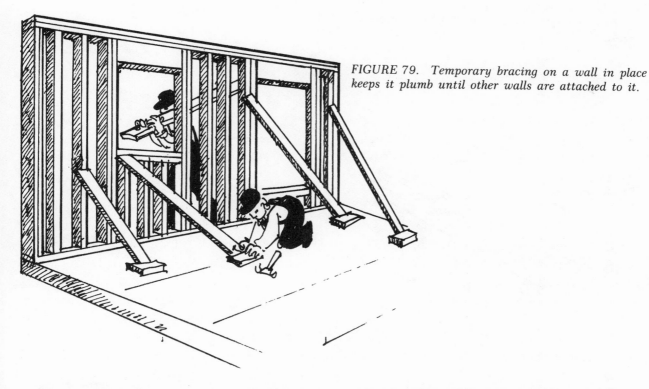

FIGURE 79. *Temporary bracing on a wall in place keeps it plumb until other walls are attached to it.*

in place, it will stay; also, the inner wall (to come) will hold it. Cut the fiberglass with a utility knife, pressing it down with a board and using the board as a straightedge or at least as a general guide.

Then in a similar fashion you build the inner walls. Before you put them into position, parallel to the outside walls, it is a good idea to run any plumbing pipes, electrical cables, and ducts in the space between the two walls (see Chapter 14). Having framed and insulated the inner walls, install them 4 inches from the outer walls and secure them to the floor in the same manner as the first. The only difference is the headers. Since these inner walls are not supporting ones, there do not have to be any headers as such over door and window openings. It is simply a matter of putting 2 x 4s on top of the rough openings at the same level as the bottom of the headers on the outer wall (Figure 80). Short studs are inserted between the top plate and header 2 x 4, and between the sill 2 x 4 of the window opening and the floor plate.

Tie the inner and outer walls together by nailing 1 x 6 boards at right angles to the walls every 4 feet or so (Figure 81A, B). There is no need to seal the top of the walls; in fact, they should be open to allow any moisture that finds its

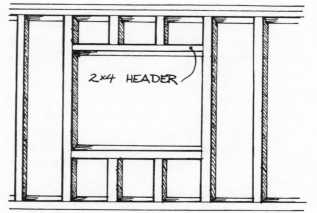

FIGURE 80. *A single 2 x 4 header with short cripple studs is used on the inner wall, which is not supporting anything.*

55

The Uppity Parts

FIGURE 81. *Two views of connectors to keep walls at the right interval. A, top view; B, vertical view.*

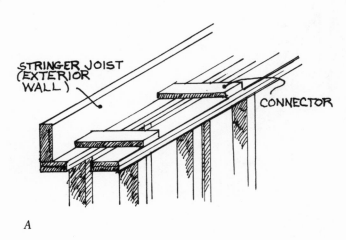

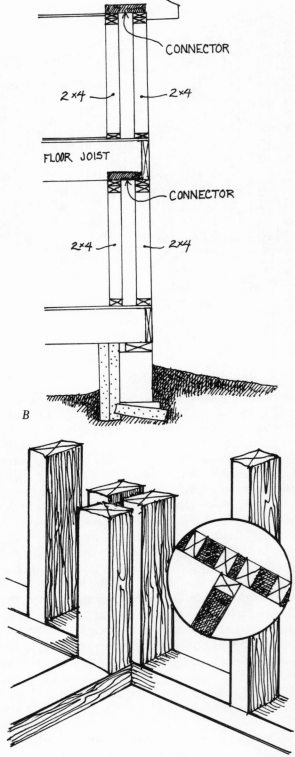

way into the wall cavity to escape. Where there is to be a stringer joist for a second floor on top of the wall, cut the connecting boards short. This is only one method for connecting walls; you can also use steel strapping. The second-floor joists or roof rafters, toenailed to the wall top plates, will aid in tying the walls together. In fact, the connecting boards may be necessary only to position the walls temporarily, until the joists and/or rafters are installed.

Interior partitions in a superinsulated addition do not have to be doubled. There is just one trick to putting up interior partitions: there must be a nailing surface where the partition wall meets the interior wall of the superinsulated wall. It is simple; just add two 2 x 4s where the partition wall butts up against the inside portion of the exterior wall. Make them 2 inches apart (Figure 82). Another method is to install a 2 x 6 so that the end stud of the partition wall butts up against the wide side of the 2 x 6 (Figure 83).

A partition wall not only divides the interior space into rooms, but if the addition is big enough, acts as a support for second-floor joists, just as the basement beam did for the first-floor joists. Such

FIGURE 82. *Extra studs are set in a wall so that a partition wall will butt up against them, leaving nailing surfaces for inside wall finishes.*

FIGURE 84. *Three ways to sheath: plywood and matched (tongued and grooved) boards, horizontal or diagonal.*

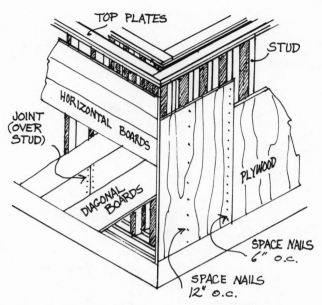

walls are built on the floor and raised into place, with the upper top plate put on last, after the wall is in position, and overlapping the lower top plates of the exterior walls.

Before you go on to build a second floor, or set rafters for a roof, it is best to install sheathing, which strengthens the entire structure and helps assure its staying plumb and level. Temporary wall bracing should remain until the roof rafters, ceiling joists, or a second floor is installed.

There are several different kinds of sheathing, but we feel matched boards (tongued and grooved), 1 x 6 or 1 x 8, or ½-inch plywood should be used (Figure 84). A third material, ⅝-inch plywood that is grooved to make it look like spaced vertical boards, can also be used, and is both sheathing and siding (finish material on the outside). Be warned that this type of siding is *plywood*, and therefore may be subject to delaminating if it is not extremely well protected.

Boards can easily be installed by one person. They are nominal 1 x 6s or 1 x 8s (meaning ¾ inch thick and 5½ and 7½ inches wide, respectively), and can be installed horizontally or diagonally. Use 8d nails, two in each stud crossing. Use three nails on diagonal boards, and set the diagonal at 45 degrees.

Sometimes the boards are not straight and take some oomph to butt one up against another. Here's a trick: if one board is curved and needs forcing to make it butt properly against its neighbor, use two nails at a time, driving them down at an angle through the tongue along the top of the board. The two nails add driving and holding power. Then face-nail in the normal manner.

A corner brace may be necessary with horizontal boards. This is a 1 x 4 or 1 x 6 set at each corner (see Figure 72). It is let in: the studs are notched to accommodate it, so that its face is even with the face of the studs.

Sheathing helps tie the structure together, particularly diagonal boards or plywood. The sill that sits on the foundation has been indented ¾ inch in from the foundation's outer face, to make room for the sheathing. When the sheathing is nailed to the wall, it not only will span the studs but also will reach down to the sill as well, including the perimeter floor joists (stringer and header joists), thus tying sill, floor, and wall into one unit.

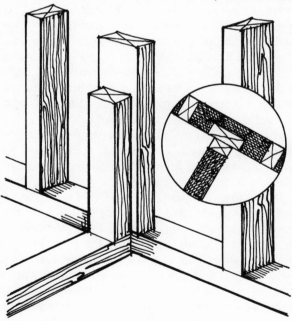

FIGURE 83. *Another treatment for a partition wall, using a 2 x 6 the long way.*

The Uppity Parts

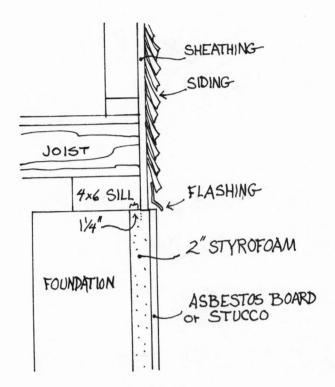

FIGURE 85. *The sill overhangs the foundation to accommodate rigid insulation on the outside of the foundation.*

If, however, insulation is put on the outside of the foundation, the sill is placed so that it overhangs the foundation to accommodate the insulation (Figure 85). If the insulation is 2 inches thick, the overhang is 1¼ inches. If the insulation is 1 inch thick, the overhang is ¼ inch. These overhangs accommodate the sheathing, which is ¾ inch thick, and are required only when insulation is on the outside of the foundation.

Plywood is also good sheathing, but it is pretty tough for one person to put up alone. If there is no outside insulation, and the sill is indented ¾ inch from the foundation, and you use ½-inch plywood, don't worry about the ¼-inch shelf remaining on the foundation. It will be covered by the siding.

Use ½-inch exterior-grade plywood. Plyscore is good. Plywood sheets, 4 by 8 feet, should be applied vertically, using 8d nails spaced 6 inches at the edges and 12 inches on intermediate studs. Set each sheet on the lip of the foundation, spacing sheets 1/16 to 1/8 inch to allow for expansion and contraction due to moisture content. Because the sill, floor joists, and wall measure considerably more than 8 feet from top to bottom, the top of the plywood will come below the top of the wall. This is good, because when subsequent sheets are installed on top of the first row, they will extend up onto the second-story wall and tie it into the first floor very nicely. When nailing plywood sheets, do not line up the joints.

With the first-floor walls and partitions up, and sheathing partially done, it's time to start the second floor. It's the same as the first; in addition, the joists not only are supports for the second floor, but also are supports for the first-floor ceiling. The joists go in the same direction as the first-floor joists, because they cover the same span and are supported in the center by load-bearing partitions.

The perimeter joists are set first: stringer and header joists, toenailed on the top plate of the walls with 10d nails and face-nailed at corners. Intermediate joists are set on 24-inch centers and overlap on the interior partition just as first-floor joists overlapped a center beam (see Figure 46). They are toenailed to the top plates of the walls with 10d nails and spiked together with 16d nails at the overlaps. These second-floor joists also are attached to the existing house as the first-floor joists did.

The second-story floor is built with proper openings (stairwell, chimney, fireplace, and so on) and the second-floor walls built and erected as the first-floor ones were.

If you plan an overhang (and this could be on the first or the second floor), it is best to have the overhang at right angles to the floor joists, so all that has to be done is to extend the joists beyond the outer walls. The exterior wall is overlapped by 10½ inches, and the joists are connected with a header joist (Figure 86). When the overhang is parallel to the length of the joists, as perhaps for a bay or overhanging window, the joists stop 6 feet from the outer part of the exterior wall and this last joist is doubled (Figure 87). Then joists 6 feet by 10½ inches are installed from the doubled

FIGURE 86. *Second-floor joists are simply cantilevered over the wall to create a second-floor overhang.*

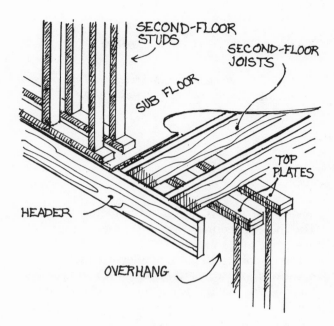

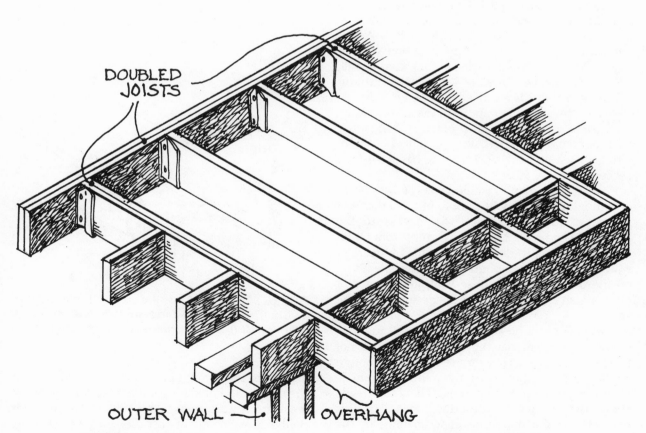

FIGURE 87. *When joists are parallel to the wall, the overhanging joists must be arranged so that they are at right angles to the wall. They are attached to a doubled joist.*

FIGURE 88. *When a partition wall parallels joists, a 2 x 6 nailer for the ceiling is nailed on top of the top plate.*

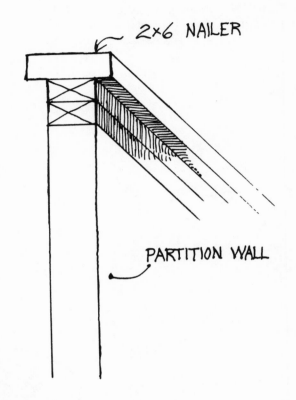

joists (use joist hangers), overlapping the wall. The header connecting the joists brings this overhang to 12 inches.

It is important to put up nailers for interior wall coverings (plasterboard or paneling) and for ceilings where ceiling joists meet walls. Where a partition wall parallels the joists, put two joists about 2 inches apart, with half of each bearing on the 2 x 4 top plate. This leaves ¾ inch on either side of the partition for nailing. A less expensive way to do it is to nail a 2 x 6 on top of the partition wall; there will be, or should be, no joists to interfere with this 2 x 6 nailer (Figure 88). Where joists are at right angles to the partition wall, or where they are at right angles to the exterior wall, there are already nailers—the joists themselves—so there is no need to add any.

Where the joists are parallel to the exterior wall, an extra joist can be installed just inside the inner wall, or a series of 2 x 4 sleepers nailed between stringer joist and the next one, at 16- or 24-inch centers (Figure 89).

Having framed and installed one (or two) floors of your addition, you will want to put in a doorway in the existing house wall for access to the addition. The process is little different from building any opening in the addition walls for doors or windows. Remove enough siding and sheathing on the outside of the house wall, and plaster or other finish material on the inside to expose the studs.

If the door will be standard width (30 inches), chances are good that the ceiling inside will not need temporary shoring up. Remove just enough studs (no more than two) and reinstall them to form the proper rough opening. Then add inside studs and header, as described above in this chapter, for the matching door opening in the addition wall. If the opening is wider than the standard, you should shore up the ceiling while

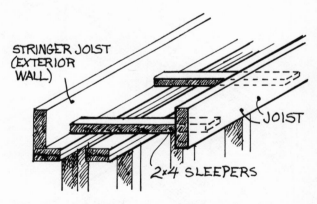

FIGURE 89. *Where joists are parallel to the exterior wall, 2 x 4 sleepers connect stringer joist and first interior joist to create a nailing surface for the ceiling.*

you make the rough opening. Set a 2 x 4 against the ceiling and hold it up with several temporary 2 x 4 posts. A wide opening must have a big enough doubled header to support the second-floor walls and other parts of the house above it (see the table on page 53).

chapter 7

Top Priority

Roof Framing and Sheathing

With two full stories on your addition, you are ready for the roof (Figure 90). A roof is made up of roof rafters, a ridgeboard, collar beams, end studs, and sheathing. Important to the structure are ceiling joists; even though they are not actually part of the roof, they contribute significantly to the roof's integrity and strength by helping to hold the rafters in place.

There are five basic kinds of roofs: gable, with a top ridge and a roof slanting down on either side of the ridge; shed, with the slant going only one way; hip, with the roof slanting down to all four sides of the structure; gambrel, with a lower steep slope and a higher shallow slope; and flat, which is dead flat or, more commonly, with a slight slope. On an addition, a gable roof may have only one exposed gable (the other end being attached to the house), or a hip roof may slope down to only three sides instead of four.

Let's start with the pitched gable roof, as being the most typical for an addition attached to the existing wall of a house.

GABLE ROOF

Pitch is the slant or slope of the roof and is determined by the rise, the vertical distance from the top of the wall top plate to the top of the ridgeboard. The span is the entire width of the roof, the run is half the span, from top plate to the midpoint of the span, directly below the ridgeboard (Figure 91). It is important to understand these terms, for their dimensions will determine angle of pitch for the roof and the size and length of the rafters.

What angle to pitch the roof? This is the first decision to make. Pitch is designed to throw off snow and to shed rain; the steeper the pitch, the more snow will be thrown off and the more warmth from the sun can be absorbed. Pitch may also be determined by the design of the addition.

An easy pitch is 45 degrees; it means that every cut in the roof elements would be made at a 45-degree angle. Suppose you decide on this angle. The addition is 30 feet wide, which means the span is 30 feet and the run is 15 feet. This means that the rise will also be 15 feet, because a 45-degree pitch rises 12 inches for each 12 inches of run. The pitch is also denoted as 12-in-12.

But this is too steep and too high for a roof on top of two full stories. It is best for a roof on a one-story addition; the area enclosed by the roof can be used as living space, with the first-floor ceiling forming the floor of this space, as in a Cape Cod–style house.

For a two-story addition, let's use a gentler angle of pitch and less of a rise, or 10 feet high.

61

Top Priority

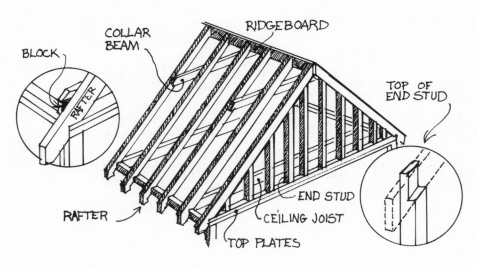

FIGURE 90. The anatomy of a roof.

With a 15-foot run and a 30-degree angle of pitch, this is an 8-inch rise for each 12 inches of run.

First you must determine the size of lumber to use for the rafters. Since they are set at an angle, they don't have to be as big as floor joists. The table shows you the size of rafter for the length of the run. To be doubly sure you have the right size, check your local building code.

Next, you determine the length of the rafters, which will be longer than the run, and the angle to cut each so that it will butt against the ridgeboard snugly and sit firmly on the top plate of

Rafter Size	Center-to-Center Spacing (ins.)	Run (ft.)	Span (ft.)
2 x 4	12	8	16
	16	7	14
	24	6	12
2 x 6	12	12	24
	16	10	20
	24	9	18
2 x 8	12	16	32
	16	14	28
	24	12	24
2 x 10	12	20	40
	16	18	36
	24	15	30
2 x 12	12	24	48
	16	21	42
	24	18	36

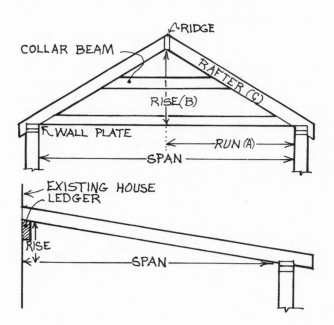

FIGURE 91. A gabled (two slopes) and a shed (one slope) roof.

the walls. For the length, you use our old friend, the Pythagorean formula: $A^2 + B^2 = C^2$. A is the run (15 feet), B is the rise (10 feet), and C is the rafter, which is the hypotenuse of the right triangle (see Figure 91). Square the run (15 x 15 = 225), add it to the square of the rise (10 x 10 = 100), and you get 325. Take the square root of 325 and you get 18 (18 x 18 = 324, close enough). So the basic rafter length is 18 feet.

But don't cut the rafter yet! You must add enough to its length so that you have an eave overhang, which is determined by the style and design of your addition, and by the amount needed to control sunlight and heat (see Chapter 2). An overhang can be any length from 6 inches to 3 feet. A narrow one is traditional; a wide one

Roof Framing and Sheathing

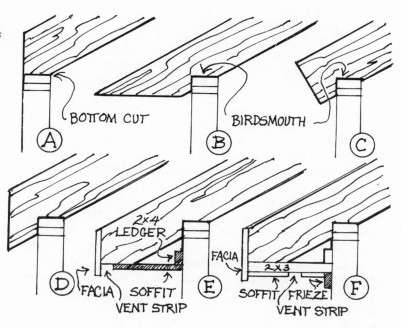

FIGURE 92. *Various types of overhangs of a roof and ways to box in the eaves.*

contemporary. A wide overhang is not recommended in climates with heavy snow and cold winters except on south-facing walls, where solar heat is a consideration. And no overhang at all isn't recommended under any circumstances because without an overhang you can run into problems: water dripping over the edge of the roof; inadequate and overfilled gutters; inadequate ventilation of the attic; and ice buildup and resulting leaks. Figure 92 shows various types of rafter, overhang, and eave treatments; details about overhangs can be found in Chapter 12.

Another way to determine the precise length of the rafters is to make a mock-up, full-sized frame, with run, rise, and overhang.

Once you have the rafter length set, including the overhang, you can figure the angles for cutting the top end that butts against the ridgeboard and the notch that fits over the wall top plate. To do this you use the framing square, one of the most useful tools you can buy. It is a large, L-shaped square, with one arm 24 inches long, called the body, and a shorter arm, 16 inches long, called the tongue. Each arm is marked in inches and is calibrated to indicate the length of the rafter and its angles. By the way, you can use the framing square to confirm the rafter length: with inches for feet, lining up the rise (10 inches) on the tongue and the run (15 inches) on the body, you will find the distance between the two points is 18 (Figure 93).

Put your rafter on the floor or ground at the angle of the roof pitch. Then place the framing square on it with the body horizontal and the tongue pointing down. Measure off the 10 inches of the rise. The line along the tongue is the line on which to cut the top end of the rafter. Measure up from the eave end the amount you've allocated to the overhang; plate the framing square in the same position along the rafter but with the tongue as high as the doubled top wall place is thick. The small right angle formed by the two arms of the framing square gives you the dimensions of the notch, or bird's mouth, that you cut to fit over the doubled top wall plate (see Figure 92). You can make the cuts at both ends of one rafter and, after you find that it fits (and it should if you did it right), you can use it as a template for the others.

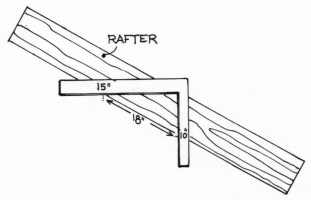

FIGURE 93. *How a framing square will determine the length of a rafter according to its pitch.*

Measure the length of the ridgeboard (equal to the length of your addition) and cut it. Then make sure the ceiling joists are in place. They are laid in the same way as floor joists but can be a size smaller because they hold up less weight (ceiling plus insulation). There are also no stringer and header joists; the top plates of the walls serve in their stead. The first joist is installed 3½ inches from the outside end, just inside the wall top plate. The rest of the joists are toenailed to the top plates of the addition walls with three 8d nails, two on one side and one on the other. The outside ends of the joists are cut at an angle to match the roof angle. Nail a sheet of plywood to the ceiling joists to serve as a catwalk while you work on the rafters. You will begin with the gable-end ones.

Install on the end top plate the gable-end studs, which must be cut to different lengths to accommodate the upward pitch of the rafters, and with notches at their top ends cut at an angle matching the pitch (see Figure 90).

When all the end studs are in place, you put on the end rafters, toenailing them to the studs and top plate with 8d or 10d nails. Leave a ¾-inch gap at the top where the two rafters meet, for the ridgeboard. Make sure this pair of end rafters and the end studs on which they rest are plumb. Then insert the ridgeboard between them, rest it at the other end on a cleat nailed in position on the existing house, and face-nail it into the ends of the rafters. Use 10d nails.

Work down the length of your addition, installing each pair of rafters one pair at a time, rather than doing all the rafters on one side first. This keeps the ridgeboard straight. If you haven't been able to get a ridgeboard long enough, make one in two pieces, butting them where a pair of rafters meet. Face-nail each rafter to its matching ceiling joist with four 12d nails.

Any openings for chimneys and skylights must be made as the rafters are installed, using the same technique as that for making openings in floors: double the rafters on each side of the opening and put in doubled connecting headers. Joists and rafters must be 2 inches away from any masonry.

All this is tough to do alone and without previous experience. But if you follow the order of procedure as described here and have at least one helper, you should be able to frame up your roof without too much difficulty.

When you reach the other end of the addition, which butts against the existing house, you can nail the rafters directly through the house sheathing into the existing studs. But if you want a really sound attachment, it is better to set the rafters at this end with end studs, in the same manner as at the outside end.

To keep the rafters from spreading (there is heavy pressure at their ends), install collar beams, or collar ties, connecting them, no farther down the rise from the ridgeboard than one-third (see Figure 90). The rafter pairs form upside-down Vs; the ties make each pair a right-side-up A. If the area above the ceiling joists and below the roof is to be used as living space, Cape Cod–style, the collar ties are put as high as 8 feet up the rise and serve as ceiling joists for the attic above, in the peak of the roof. If used this way, they must connect each pair of rafters. If they are simply collar ties, they can connect every third or fourth pair. A collar tie can be a 1 x 6, 1 x 8, 2 x 6, or 2 x 8.

Treatment of the overhang at the eaves is described in detail in Chapter 12; treatment of an overhang at the rake, or gable end, of the addition may require modification of the end rafters if it is 12 inches or more. The end rafters are eliminated, and the next pair of rafters in from the gable is doubled. Gable-end studs are cut at the same

Roof Framing and Sheathing

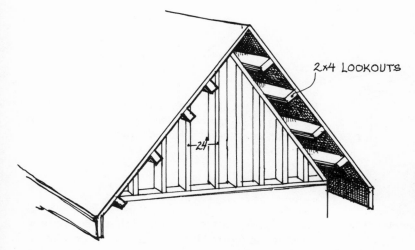

FIGURE 94. *For an overhanging rake, gable studs are cut at the angle of the roof pitch, topped with a plate, and 2 x 4 lookouts are cantilevered over them. The gable rafter is eliminated.*

angle as the rafter pitch, installed, and covered with a 2 x 4 plate so that they are 3½ inches below the high side of the rafters (Figure 94). Then 2 x 4 lookouts are extended from the doubled rafters over the 2 x 4 plate, on which they rest, and allowed to cantilever beyond the gable wall. They literally "look out" over the wall—hence their name. The lookouts are attached to the doubled rafters by joist hangers, or they can be toenailed to them. Nailing blocks of 2 x 4s or 2 x 6s are nailed flat on the plate between the lookouts, overhanging it by an inch or so, to act as a nailing surface for the soffit board, the bottom of the overhang. More about soffits and the rake end in Chapter 12.

SHED ROOF

A shed roof is considerably easier to make than a gable roof, because it has only one pitch (see Figure 91), from the high side of the addition to the low side, with the high side butting against the existing house. It may pose problems by interfering with windows in the existing house. In some cases these windows can be made smaller so that the shed roof can start higher.

Rafters that butt up against the house wall can sit on a ledger of the same size as the rafters (see Figure 91). They can be butted against the ledger and secured with joist hangers, or can be set directly on the ledger and toenailed to the existing house.

Structures such as ledgers that are attached to the existing house can be nailed directly onto the siding, through siding and sheathing and into studs. But it is better to remove the siding where these attachments come, for two reasons: you get a vertical, flat surface to nail to, and it is easier to find the studs.

To mark the rafter in a shed roof for cutting, set it in place, with the proper overhang. Prop it up in position so that its end just touches the house wall. You will have to nail it. Then take a short straightedge and butt it up against the wall, with its drawing edge on the side of the rafter. Make a mark and cut it. Then cut the bird's mouth at the other end, where the rafter sits on the opposite wall, noting the overhang and the proper place for the cut. You can keep the rafter end square, so that the overhang has the same slant as the roof, or cut it so that it is plumb (vertical). To do this, use a bevel, or adjustable, square, equipped with a wing nut for tightening the bevel in the right position. Set the bevel at the mark of the upper end of the rafter and transfer the angle to the other end. After cutting the upper end and the bird's mouth, check for fit. If it fits properly, use this rafter as a template for the others.

Sometimes a shed roof will attach to the pitched roof of the existing house (Figure 95). Putting two roofs at different pitches like this is to be avoided, because the juncture of the two pitches is a potential leak spot, and a chronic one at that.

But if it has to be done, the rafter spacing must correspond to that of the rafters in the existing house. The new rafters are cut to butt against the existing rafters. They are toenailed through roofing and sheathing and into the existing ones. Another way is to remove portions of the roofing and sheathing and nail the new rafters directly against the existing ones.

Top Priority

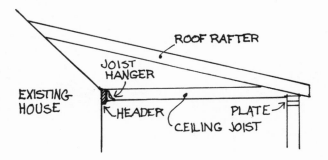

FIGURE 95. *A shed roof running into the slope of an existing roof should be avoided, because the change of pitch can cause leaks where the two roofs meet.*

SPECIAL ROOFS

We call hip and gambrel roofs special, not because they are uncommon but because they are the very devil to build. Gambrel roofs aren't really that difficult, but hip roofs are.

A hip roof is one that slopes down on all four sides of a structure (or on three, for an addition), and has a short ridge, a long ridge, or no ridge at all (Figure 96). In the three-sided hip roof, the ridge could parallel the wall of the existing house or it could stick out at right angles to it.

Generally a hip roof is used according to the style of the existing building. It can keep the roof from being too high, preventing the soaring look of, say, a French provincial farmhouse. In addition, a hip roof has stability, plus less tendency to rack (move from the vertical).

A hip roof consists of common rafters (full length) and hip rafters, from one end of the ridge to the corner of the addition. The pitch of the hip (the short end roof) should be the same as the pitch of the common rafters; otherwise the end cut must be figured out with the rafters square so that they will butt properly against the ridgeboard.

Once all common rafters and the hip rafter are installed, butting against the ridgeboard, jack rafters (called hip jacks) are cut to extend from the top plate of the wall to the hip rafters. The hip rafters must be 2 inches deeper than the common rafters so that the jacks, cut at an angle, can make full contact on them. Nail jacks to hip rafters with 10d nails, and toenail them to the top plate with 10d nails.

The most difficult part in making a hip roof is cutting the special rafters. The jack-rafter ends have to be cut at two angles: one to match the pitch of the roof, the other to match the angle of the hip rafters. Of course, both angles are cut at the same time, which makes the work even more scary.

Where two gabled roofs intersect—for exam-

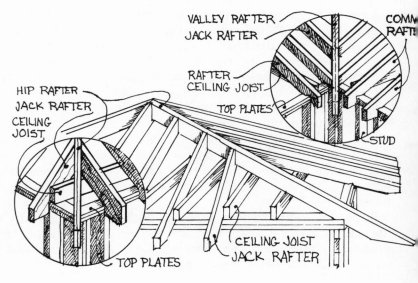

FIGURE 96. *Special treatments of joists are needed to create a hip roof or to make a valley.*

Roof Framing and Sheathing

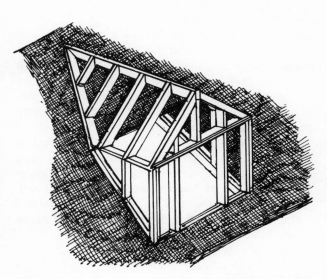

FIGURE 97. *A small gabled roof tied into another roof. Nail plates directly on the roof as a base for side wall studs and jack rafters.*

ple, when the gabled roof of an addition in an *L* shape meets the gabled roof of the existing house—you use a roofing technique similar to that used in making a hip roof, only in reverse. Here instead of a hip, going up and out, you will have a valley, going down and in. You will need valley rafters. These rafters, doubled, extend from the inside corner of the *L* to the ridgeboard. Then short valley jacks, 2 inches less deep than the valley rafters, are face-nailed to the ridgeboard and toenailed to the doubled valley rafters with 10d nails (see Figure 96). When a small gabled roof, like that of an "A," or eye dormer, is tied into another roof, plates are nailed directly on the roof as a basis for both side wall studs and short jack rafters that connect the plates to the ridgeboard (Figure 97).

A gambrel roof, seen on Dutch Colonial houses, is generally built from the top of a first-story wall and gives considerably more room on the second floor for living space than a gable roof on a typical Cape Cod–style house. Barns have gambrel roofs for the same reason: lots of storage space on the second floor.

The steep slope of a gambrel roof is 60 to 70 percent of the total rise, and the shallow slope, from there to the ridgeboard, takes up the rest of the rise (Figure 98). Another advantage of the gambrel roof is that it lends itself to solar applications: if the steep lower part faces south, it can be a full window wall to allow the sun to shine in. This is particularly useful if the first floor is open to the roof.

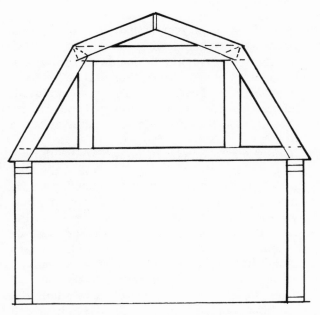

FIGURE 98. *Gambrel roof, with two slopes of different degrees on each side of the ridgeboard.*

Top Priority

The steep rafters can be smaller than those in a regular roof. They are butted onto the top plate of the first-story wall, and their tops are connected with collar beams, which are also ceiling joists. The top rafters are set on a ridgeboard and overlapped and nailed onto the steep rafters. This construction does not allow for an overhang; since one is generally needed for proper ventilation, it is better to set the second-floor joists so that they overhang the wall. Then the steep rafters are overlapped on these joists and nailed to them.

Flat roofs are discussed in Chapter 12.

All roofs except the gable and shed roofs are specialized, so the amateur really should borrow or hire a professional framer for them.

Roof Sheathing

Sheathing on a roof is like that on a wall: matched spruce boards or ½- or ⅝-inch plywood. A single worker can handle boards better than the big panels of plywood; boards should be 1 x 6s or 1 x 8s. Matched boards are tongued and grooved; end-matched boards are also tongued and grooved at the ends, so they don't have to be butted over a rafter.

One thing to watch out for when sheathing a roof is the overhangs, at the eaves and at the rake (gable) end. Because sheathing is applied before overhang details are installed, you have to know how much to overhang it. This is easy at the eaves; the sheathing overhang is determined by the length of the rafter. All you have to do is go beyond the rafter ends by 1½ inches to accommodate the facia (front of the eave) and the molding at the top of the facia. If you don't overhang the sheathing, don't worry. You can simply install facia and top trim even with the sheathing; the drip edge and roof shingles will cover them. See Chapter 12 for details.

For an overhang at the rake end, you can overhang the sheathing by up to 16 inches without excessive sagging, and it's okay to do this until the proper support is installed.

Another thing to watch out for is that the sheathing must be ¾ inch away from any masonry, such as a chimney.

Nail the boards on each rafter with two 8d nails. Use hot-dipped zinc galvanized box nails; they work as well as or better than threaded or ring-shanked nails. If boards are not end-matched, they must meet end to end over a rafter. No two boards should meet end to end over the same rafter in succession.

Plywood can be at least ¼-inch thick, but this is too thin for practical purposes, especially if rafters are 24 inches o.c. Half-inch plywood is typical and adequate, ⅝-inch more than enough. Lay plywood with its grain perpendicular to the rafters so that it won't sag. Stagger butt joints so they won't meet on the same rafter. Nail plywood with 8d galvanized box nails, 6 inches apart on edges and 12 inches apart on intermediate rafters.

On additions with cathedral ceilings and exposed beams, the sheathing is actually a kind of decking, nominally 2 inches thick, such as 2 x 6s, sometimes tongued and grooved. It is designed to span rafters farther apart than 24 inches. When beams are exposed, the decking acts as a base for the roof as well as forming the ceiling inside. Insulation is applied on top of the decking and the roofing material on top of that. In terms of superinsulation, as we have seen in Chapter 3, a cathedral ceiling presents a problem.

At hips, valleys, and ridges when there is no ridge vent, the sheathing should be as tightly butted as possible. Don't worry if the sheathing sections don't touch at all points along the entire thickness; any gaps will be covered by roof material. At valleys, the tightly butted sheathing at the top will form the proper back for asphalt or metal flashing under interlaced asphalt shingles.

VENTILATION

In building a roof ventilation should be planned for; the attic and roof area is an important one in the ventilation system of the addition. The ratio of ventilating space to attic floor area is 2 to 300; one square foot of space for air to enter the attic and one square foot for air to exit for each 300 square feet of attic area. So if your attic floor area is 900 square feet, you need 3 square feet of ventilating space, or a vent 18 by 24 inches, at each end of the attic, in the gables (Figure 99). If the gable end of your addition butts up against the existing house, obviously you won't have a vent at that end. You should put a vent at the other, open gable end, or you can use a ridge vent. When the addition attic roof butts against the existing attic roof, an opening can be made in the wall or roof to allow air to circulate through the two attics.

Another good place for ventilation is at the eaves, important in an addition with only three exposed sides and in hip roofs. Equally good is a ridge vent on a long gable roof or hip roof (Figure 100), which permits a more than adequate flow of ventilating air. To install a ridge vent, stop the sheathing at least 2 inches on either side of the ridgeboard. The vent will be nailed later, through roofing and sheathing, just below this opening (see Chapter 12). The roof shingles will stop at the opening, and the vent will cover them. Eave ventilation is discussed in Chapter 11.

On flat roofs and pitched roofs forming a cathe-

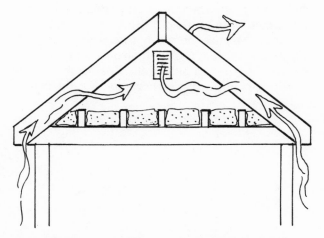

FIGURE 99. *Ventilators are good at the gable ends, better in conjunction with soffit vents (those in the underpart of the overhang).*

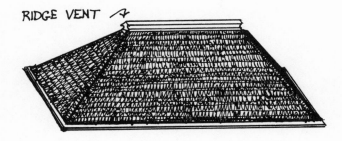

FIGURE 100. *A hip roof often has a ridge long enough for a ridge vent. Added to soffit vents, a ridge vent will provide more than adequate ventilation.*

dral ceiling, in which roof and ceiling joists are combined, a minimum of 1½ inches is required as ventilating space between insulation and roof sheathing. In a flat roof with no overhang, ventilation can be in the side wall; this arrangement, however, is not very weatherproof. For the cathedral ceiling, the space between insulation and roof sheathing not only must be open but must have flowing air. Thus you should install both eave vents and a ridge vent (see Figure 31). Remember—there is no such thing as overventilating, but there is definitely such a thing as underventilating.

chapter 8

On the Up and Up

Dormers and Second Floors

Rather than build an addition from the ground up, as described in the preceding chapters, you may decide to put on a dormer or a full second floor. Is your existing house strong enough to support this kind of addition?

If a house can support a roof and roof rafters, it can support a dormer. A one-story house can also support a full second floor; there is no difference in construction between a one-story and a two-story house. However, a one-story house may have inadequate joists holding up the ceiling because they were designed for that purpose, not to hold up another floor. So, as we will see later in this chapter, you should add heavier joists to support a second-story floor and walls. Those walls in turn will support a new roof.

Adding up — dormers or second stories — also poses the problem of removing roofs and exposing parts of the existing house interior to the elements. So speed is important here. Remove and replace the roof as fast as possible. If the roof opening is exposed overnight, or if an unexpected summer storm hits, be prepared. Tack a polyethylene or canvas tarpaulin to the top of the roof with a 1 x 3 nailed along the edge of the tarp so that you can roll it down to protect the house's innards. If you are building an entire second floor, you should cover those innards regularly at night or whenever you are not working on the project. It pays to work on this type of addition when a period of pleasant weather is firmly assured.

SCAFFOLDING

Next to protection of the inside of the house, safety is important when you are working on an upper level. It is best to use scaffolding instead of ladders. You can rent steel scaffolding, which will save you time, if not money.

You can also build your own scaffolding, reached by one or more ladders (Figure 101), which isn't fancy but *is* safe. Use 2 x 4s for posts. Place the front posts 6 inches or so from the house and the rear posts 3 feet. They can be 8 feet apart. In soft soil, support them with 2 x 6 pads or bases. Toenail the posts to these bases. A horizontal 2 x 4 brace connects the posts and ties them into the house. The brace can be nailed to the side of the house at a corner or toenailed into the siding and sheathing. It is better to nail a 2 x 4 cleat to the house vertically and nail the brace to the cleat. To keep everything from racking, 1 x 6 cross-bracing connects the posts in a X shape. Floor planks are 2 x 8s or 2 x 10s. Nail cleats to their bottoms just inside the braces so that they won't move laterally.

Make sure posts are plumb and floor planks level. They don't have to be right on the mark, but

On the Up and Up

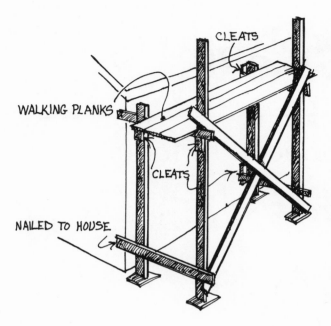

FIGURE 101. *A wood scaffolding. It is obviously important to build a scaffolding carefully and accurately, even if it is a temporary structure. You can also rent scaffolds.*

the more plumb and level they are, the safer they will be. Never step back to admire your work while you're on a scaffold. A 2 x 4 railing 36 to 48 inches high can connect the posts if they stick up beyond the floor planks.

For work higher than two stories, use 2 x 6 posts and secure them to the house at the bottom, top, and middle. Use single-length posts wherever possible. If not, butt two 2 x 6s together and secure them with 2 x 6 or plywood gussets nailed on both sides of the posts.

Dormers

A second story can be expanded by building a shed dormer (Figure 102). It can go the full length of the house or can be set in. It can go to the outer wall or stop short of it. These choices depend upon how much room you want or need.

We will describe a dormer that is set in on both sides and is a foot or two short of the outer wall. Such a dormer makes the roofline of the house more attractive.

The general outlines of the dormer are shown in Figure 103. The dormer roof will have less of a pitch than the existing roof. It should have at least 3 inches of rise (vertical height) to each 12 inches of run (from the ridge horizontally to the wall) to allow installation of roof shingles. If the pitch is any less, roll roofing, which is less desirable for a roof than shingles, is required.

The pitch is determined by the height of the roof (Figure 103) at the ridge line (A) and the height of the new outer wall of the dormer (B). The wall should be at least 6½ feet high or higher if possible. If it is 6½ feet high and the ridge 10 feet high, the rise is 3½ feet. If the distance from the ridge line to the wall is 12 feet, you have a pitch of 3½-in-12, steep enough for roof shingles.

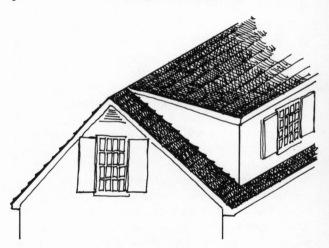

FIGURE 102. *This shed dormer is indented both from the gable wall and from the back wall. The indentation is good style.*

When there is no subfloor on the second-floor or attic level, install insulation or add to it before putting down a subfloor. You can't very well work without one. Make sure electrical cables and plumbing are installed where necessary before the subfloor is laid. The best subfloor is ⅝-inch plywood. If large sheets cannot be brought up to the attic level, use 1 x 6 or 1 x 8 matched spruce boards.

Dormers and Second Floors

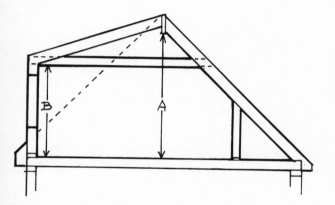

FIGURE 103. *The outline of a shed dormer. The height of the dormer to the ridgeboard (A) and the height of the outer wall (B) determine the pitch of the roof.*

Why install or add insulation in the floor between two areas that are both going to be heated? When the two floor areas of a house are fully insulated, they will retain more of the heat that is introduced to them. If the heating system is zoned, and one area is kept cool, there will be less heat loss, or possibly none at all, from the warmed area to the cool area.

Before you remove the roof and rafters, you must shore up the ridge and the parts of the rafters that will remain in front of the wall. This requires only 2 x 4s or 4 x 4 posts (Figure 104) and a 2 x 4 header with 2 x 4 posts under the rafters near the eaves.

To mark the area of the roof to be cut, drop a plumb line from the third rafter in from one end of the house with the plumb bob over a mark on the floor as far in from the eave as you plan the wall (Figure 105). Do the same on the third rafter in from the opposite end. Connect these marks with a chalk line. Make vertical marks on the rafters, using a bevel square or a level. Drive a nail through the roof on the inside side of each of these rafters; that way you won't be cutting the rafters themselves. The nails will stick out through the roof. On the outside, connect the nails with a chalk line (Figure 106). Vertical lines can be made with a chalk line also: measure the distance

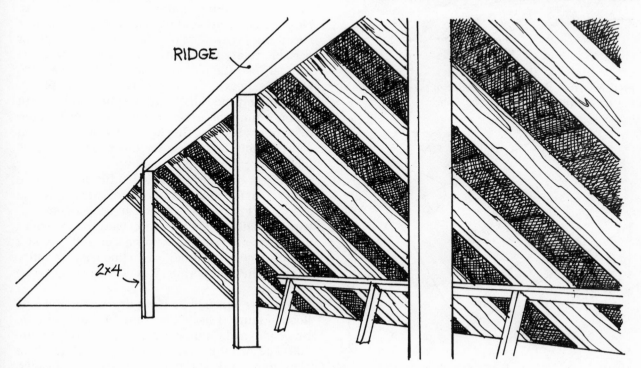

FIGURE 104. *The ridgeboard is shored temporarily by 2 x 4 posts. Rafters are supported near where they will be cut to accommodate the dormer opening.*

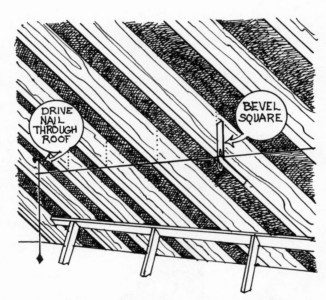

FIGURE 105. *Marking the inside of the dormer opening as a guide for cutting the roof.*

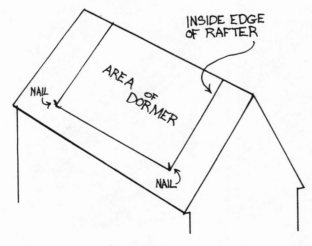

FIGURE 106. *Snap chalk lines on the outside of the roof to indicate where shingles and sheathing are to be cut.*

from the nails to the ends of the roof on either side and transfer this distance to the ridge. Now you have an outline on the roof of the dormer's dimensions.

You should build a stud wall for the outer wall of the dormer. You can build it right on the floor of the attic so that it is ready to be put into place when the roof and the existing rafters are removed. It can have studs on 24-inch centers and openings for any windows.

Now you are ready for the surgery. Be sure you have tacked the protective tarp or plastic sheeting to the ridge. Remove the ridge shingles and the roof shingles inside your chalk lines. You can cut them with a utility knife right along the lines. Next, remove any roofing felt. Then cut the roof sheathing plywood or boards with a saber saw or portable rotary-blade saw. Cut carefully because you can use the old sheathing for the new roof. Cut the rafters at the marks you made earlier and pry them off the ridgeboard. Chances are the cut rafters cannot be used for either ceiling joists or new rafters because they will be too short, but you can probably find a use for them. So remove all nails and carefully put them away.

Connect the cut rafters with a header, one size larger than the rafters. If the rafters are 2 x 8s, use a 2 x 10 (Figure 107). Nail it to each rafter with two 16d nails, lining up its top with the top edges of the rafters. Nail a second header to it, keeping the top edge of the inner one in line with the rafter angle. The temporary 2 x 4 braces can be removed.

You have finished building the wall on the attic floor (Figure 108). Now raise it into position. Brace it to make sure it is plumb and level (Figure 109) and nail it in position. Nail through the floor plate into the joists and toenail into the header. Build and install a second wall for superinsulation (Figure 110).

Dormers and Second Floors

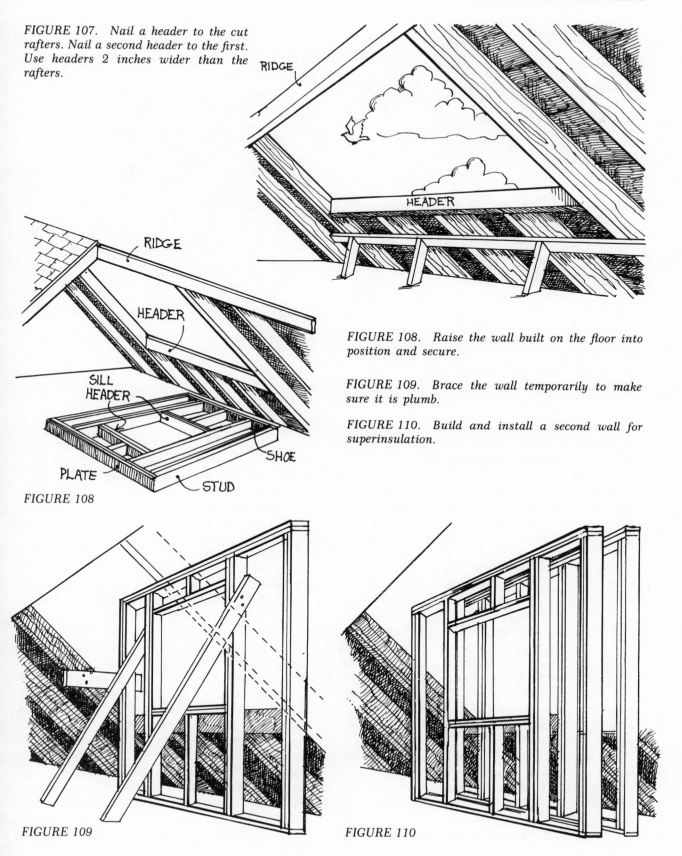

FIGURE 107. Nail a header to the cut rafters. Nail a second header to the first. Use headers 2 inches wider than the rafters.

FIGURE 108. Raise the wall built on the floor into position and secure.

FIGURE 109. Brace the wall temporarily to make sure it is plumb.

FIGURE 110. Build and install a second wall for superinsulation.

On the Up and Up

FIGURE 111. *Ceiling joists go from top of outer wall to existing rafters on the opposite side of attic. Joist spacing will have to correspond with the spacing of existing rafters.*

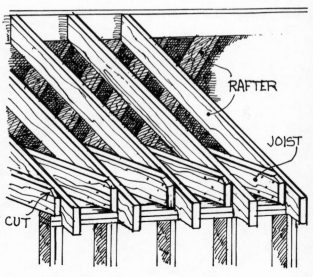

FIGURE 112. *Cut rafters to go from outer wall to ridgeboard.*

Cut ceiling joists to go from the new outer wall to the existing rafters on the opposite side (Figure 111). To facilitate positioning the joists, nail a temporary 1 x 3 cleat for them to rest on. Use 2 x 6 or larger joists. Nail into the rafters and toenail into the top plate of the new wall with 16d nails. Joists at each end of the dormer should be indented 3 inches to accommodate the end rafters and end studs.

Now cut the new rafters (Figure 112). Here is how to do it for the dormer: set a rafter (2 x 8 or larger) against the ridgeboard, with its other end on the new wall (Figure 113 A). Support it against the ridgeboard and line a level up beside it, with the level touching the ridgeboard. Mark this angle. Then mark the other end of the rafter for the bird's mouth and end cuts (Figure 113 B). After cutting, check for fit. If the bird's mouth fits over the wall and the top of the rafter fits squarely against the ridgeboard, use this rafter as a template for the others.

Incidentally, the new wall studs and rafters can be set on 24-inch centers, but the ceiling joists will have to be on 16-inch centers or whatever the old rafters are set on, in order for the joists to be nailed to them.

End walls are a little trickier. It is best to set the studs on 24-inch centers, but they must be set on a floor plate and their tops cut and notched to fit onto the end rafters. Toenail them to the floor plates and to the end rafters, using 12d nails. Face-nail through the ceiling joist into each stud. The entire end wall should be doubled (for superinsulation) and set on the floor, connected at the top by steel straps.

Sheathing on the sides of the dormer is nailed up to the tops of the rafters, and then the roof sheathing is applied. Of course, to get the dormer under a roof as soon as possible, you can put the roof sheathing on first.

On the opposite side of the attic, a short knee wall is installed. The height of this wall is arbitrary, but usually 4 feet is adequate (Figure 114). Drop a plumb line from the end rafter at the 4-foot-high mark, and mark the floor and the rafter. Do the same at the other end and connect the marks with a chalk line on the floor. Don't install the wall there, though. Instead, measure 11 inches from the mark toward the eaves and snap another chalk line. This is where the outer wall of the double wall will go, to allow superinsulation. The higher inner wall will be 4 inches from the outer wall. It is easiest to nail floor plates to the floor, cut studs to length and toenail them to

Dormers and Second Floors

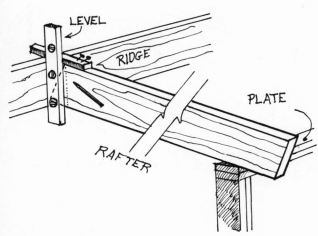

FIGURE 113 A. *Set rafter in place and secure temporarily to the ridgeboard with a 1 x 2 connector. Place the level against the ridgeboard and draw a line on the rafter to correspond with the vertical edge of the ridgeboard.*

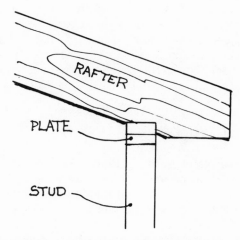

FIGURE 113 B. *Mark the other end of the rafter for the bird's mouth and cut the rafter end both plumb and level, to allow for the vertical facia and the horizontal soffit.*

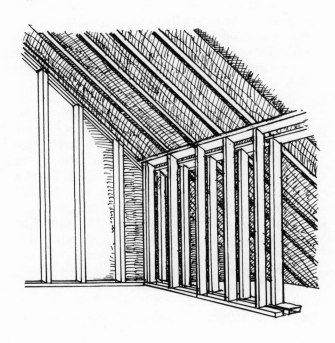

FIGURE 114. *The knee wall opposite the dormer can be any height. Make it plumb and double it for superinsulation.*

77

the floor plates, and face-nail them to each rafter. This technique requires them to be spaced the same center-to-center distance as the rafters. If you want to space them 24 inches o.c., you can nail a top plate to the rafters and toenail the studs to floor and top plates, using 12d nails.

HALF-GABLE DORMER

A half-gable dormer (Figure 115) is necessary when the space under the roof is not adequate, as in a ranch house, and you don't want to raise the entire roof. It also could work in the shallow attic of a third floor.

The main difference between an ordinary shed dormer and the half-gable dormer is in the ridgeboard, which must be replaced with a beam to hold up the wall—usually glass—that forms the top, vertical part of the half gable. A half-gable dormer can be indented, or set in, as described in the section on shed dormers, and its main wall extended to the main wall of the existing house or stopped a foot or two before reaching the main wall.

Before roofing, sheathing, and rafters are removed, the rafters of the roof that will not be removed must be shored up temporarily (Figure 116) or a high knee wall built (behind which storage can be established). If the area is big enough so that half the floor space under the roof can make an adequate room, the wall can be built under the ridge (Figure 117) and the space behind it used for storage. The wall under the ridge would be adequate to hold up the rafters and the half gable.

But if the ridge is not supported that way, it must be supported by a beam. Once everything is shored up, remove ridge shingles, roofing, sheathing, and rafters. Because the beam must be

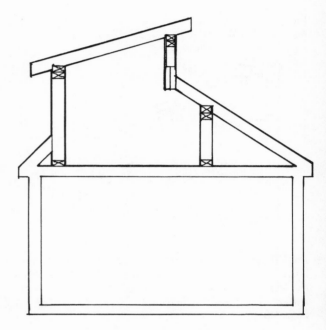

FIGURE 115. *Half-gable dormer raises both the high and the low points of the roof, allowing for a set of clerestory windows above the ridge.*

supported at each end, which means each gable wall, all the rafters must be removed except the end rafters, after all the roofing and sheathing is taken off. The end rafters must be cut to accommodate the new beam that takes the place of the ridgeboard (Figure 118). After cutting the rafters, remove the ridgeboard and replace it with doubled 2 x 12s. Posts (4 x 4s or doubled 2 x 4s) are built into the wall (Figure 118). The old rafters opposite the new room, the ones that will not be removed, will be smaller than the 2 x 12s. The bottom of the beam will line up with the bottom of each rafter, so that the beam extends above the tops of the rafters (Figures 118, 119). This will add to the

Dormers and Second Floors

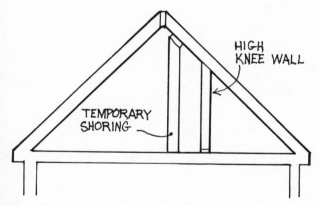

FIGURE 116. *Shore up rafters that will not be removed. If the high knee wall is close enough to the ridgeboard, the temporary shoring may not be necessary.*

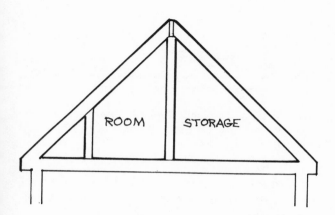

FIGURE 117. *If the attic is big enough to allow half of it to be a new room, the wall can be made under the ridgeboard.*

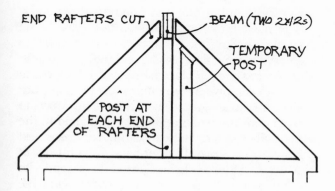

FIGURE 118. *To make the half gable, the wall under the ridgeboard must support a beam that will take the place of the ridgeboard; to do this, the rafters that will not be moved or cut must be temporarily shored.*

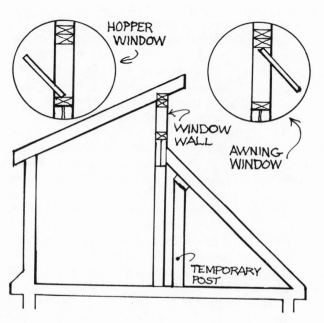

FIGURE 119. *Build the window wall on top of the new ridgebeam. It can contain hopper or awning windows; the latter are more weatherproof.*

79

clearance between the roof and the bottom of the windows in the new glass (window) wall.

Now, build the window wall to sit on top of the new ridgebeam (Figure 119). You will have to determine the type and size of windows to go into this wall. They could be awning or hopper type (Figure 119), but they should be openable to allow escape of heat during warm weather. The window wall is a simple 2 x 4 stud wall, with a floor plate that sits on the ridgebeam and a doubled top plate.

It would be helpful if the new windows, clerestory windows in effect, face south, to allow solar heating. Better yet, if the opposite wall faces south, *that* wall could be made virtually all glass, for a much greater solar gain. In either case, and even if the new walls face east and west, the windows in them should produce enough light so there would be no need for windows in each gable end.

Now it's time to put up the new main wall (usually facing toward the back of the house) and rafters. Build the wall double, as described earlier. New rafters for the half-gable dormer have considerations that differ slightly from those for a shed dormer.

If you want to indent the gable walls of the dormer, you must put in 2 or 3 rafters at each end (Figure 120). They butt against the new beam and can be splinted against the cut-off rafters at the low (opposite) side of the dormer. A dormer roof with clerestory windows necessarily has a cathedral ceiling, following the contours of the roof. It is impossible to put a flat ceiling in such a room because the clerestory windows go clear to the roof (see Figure 119). So new rafters must be cut to overhang the clerestory window wall by at least 6 inches (Figures 115, 119). This is to allow proper ventilation between insulation in the ceiling and roof sheathing (insulation and ventilation in the roof area are covered in Chapter 12).

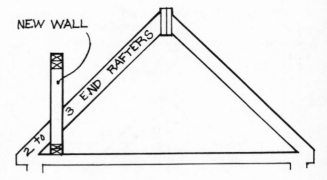

FIGURE 120. *To indent the new dormer, a new wall must be built one, two, or three rafters in from each gable end.*

Once the frame is up, all you have to contend with is sheathing, roofing, and other details discussed elsewhere in this book.

Raising Entire Roof

If you want to expand a ranch house, you can take off the entire roof and build a second story.

The most important consideration here is the floor. The ceiling joists are probably too small to support a second floor, walls, and roof on top of it, so they must be replaced. Instead of removing them, you simply have to remove the bridging and nail new heavier joists (see Chapter 5 for proper sizing) onto the existing ones. And, to add to the second-floor space, you can cantilever these new joists 12 inches over front or back or both sides of the house (Figure 121). Trying to extend the second floor over the ends of the house where the joists are parallel to the wall is another matter, not worth attempting.

Once the new floor joists and stringer and header joists, plus the floor sheathing, are put on,

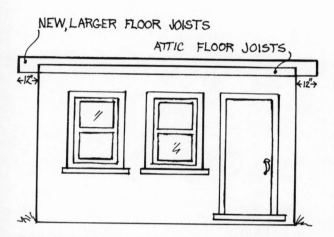

FIGURE 121. *For a new second floor, new, larger joists are installed beside the existing, smaller joists. Remove any bridging to that new joists will fit beside the old ones. The new second floor extends 12 inches over the front and back walls to create a garrison style.*

it's time to build the second floor, which is built just like the second floor of an ordinary addition (see Chapter 6).

Insulation on the first-floor ceiling may already be in place. If so, it should be removed to accommodate the new joists. It can be put back, or new insulation (6-inch unbacked fiberglass) put in its place. At the overhang, put in as much insulation as will fit. We mention this now because once the floor sheathing is installed, there is no way to put in insulation except by blowing it in, an expensive operation. It is most important to have good insulation in the overhang, to prevent floors from being cold and cold air from entering the space between ceiling and floor.

chapter 9

Ups and Downs

Stairways

If your addition is on the second floor, there has to be some way to get to it. It's as simple as that, unless you already have a finished staircase (Figure 122). Stairs and steps sound very complicated, but once you understand them, they, too, are simple.

Staircases come in all shapes and forms, prebuilt and preassembled; sometimes in kit form. All you have to do is buy one and have it installed in the proper place and position. If you want to do it yourself, you can buy the component parts, but that is impractical considering the ease of prebuilt stairs. Among prebuilt staircases are ones with platforms that allow the stairs to turn a corner.

Locating the stairs depends on your own plans, but if the stairway is to lead from the first floor to the second, the best place for it is over any existing basement stairs. The type of staircase will be determined by its location and the size of the location. A straight staircase takes up more room than one with a platform that turns a corner. And there are winder stairs: that is, with pie-shaped treads that taper from an extra-deep depth to nothing. Some building codes do not permit the use of this type of tread; even if they are allowed, the center of each tread must be at least 10 inches deep so that the stair climber, who usually steps on the center portion, will have enough tread for safe support.

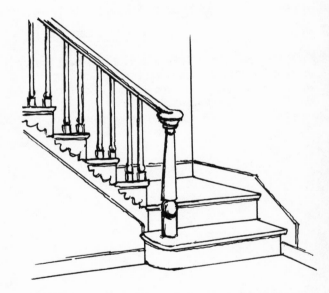

FIGURE 122. *A ready-made staircase, in kit form, is usually preassembled before installation.*

The dimensions of steps in a standard staircase follow a formula: the depth of the tread plus two risers (twice the height of a riser) should equal 25 inches. A good riser height is 7½ inches; two equal 15 inches, so the tread depth should be 10 inches. If a higher or lower riser is necessary, the depth of the tread should be adjusted to fit the 25-inch formula. This formula was designed to allow comfort and balance for the stair user.

The amount the tread overhangs the riser is about 1⅛ inches and is called a nosing, which requires a trim below to support it. In determining the height of the riser, the recommended height of 7½ inches includes the thickness of the tread. The depth of the tread runs from nosing to nosing.

Headroom is important, too. A standard staircase requires a minimum of headroom so that the stair climber won't bang his head on the ceiling or anything above the staircase. Head clearance for a standard staircase is a minimum of 6 feet 8 inches (Figure 123).

The width of the steps is a minimum 2 feet 8 inches; 3 feet is better. Thus the opening for the staircase is at least 9½ feet long and 32 to 36 inches wide. The number of steps in a prebuilt staircase can vary. Fourteen risers are standard for an 8-foot ceiling (which must include the joist depth and floor thicknesses, totaling about 105 inches). Divide 14 into 105 and you get 7½ (inches), the standard riser height. If the height is less, perhaps 13 risers would be adequate. When you have 14 risers, you have 13 treads or steps; 13 risers, 12 steps.

Don't measure or cut any holes in the floor until you have the correct dimensions of your staircase. To get the right height, measure the height of the ceiling plus the thickness of the subfloor and finish floor on the second story. If you don't know

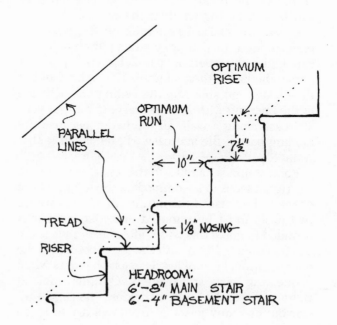

FIGURE 123. *Ideal stairway has a 7½-inch riser and a 10-inch tread. Headroom is important to prevent climbers from bumping their heads on the stairway ceiling.*

Ups and Downs

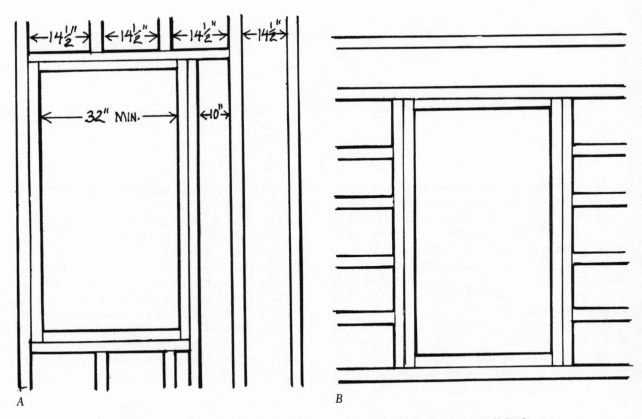

FIGURE 124. *Opening in the floor for a stairway: A, when opening parallels the joists; and B, when opening is at right angles to the joists.*

the thickness of the finish floor yet, it is not absolutely necessary to provide it. The top riser can be a little higher than the others.

Locate the staircase opening on the floor and relocate it on the ceiling, using a plumb bob. You can make the opening parallel to the joists or perpendicular to them (Figure 124 A, B). Cut out the plaster and shore up the ceiling on each side of the opening with a 2 x 4 brace. If you put your staircase along a wall, all the better because the holding part of the staircase will be nailed to that wall.

For a staircase parallel to the joists, cut them in the right places to accommodate the length of the opening. Because of the minimum 32-inch width, two joists must be removed and a stringer joist installed to make this width after header joists are put up (Figure 124 A). All the joists bordering the opening must be doubled. Use joist hangers when connecting joists and headers. Cutting a hole at right angles to the joists is trickier (Figure 124 B). Cut out as many joists as are necessary to reach the right length and install header joists on each side of the hole, connecting the cut joists. Double them, and double the short joists at each end of the opening, too.

Some staircases are made with housed carriages, which are 2 x 12 boards with notches routed out to half the boards' depth to support the treads and risers. Prebuilt staircases normally don't need carriages, except perhaps for one, in the middle of the flight. But if you build your own stairway, you must use carriages (Figure 125). There must be at least 3½ inches of solid wood from the deepest part of the notch to the back of the carriage. The thickness and width of the treads dictate how many carriages to use. Three are needed with ¾-inch treads more than 2½ feet wide, and 1½-inch treads more than 3 feet wide.

The notched carriage boards are put in place, reaching from the floor to the short header joist at the ceiling opening (Figure 126). Make sure the top of the carriage is far enough below the floor above to allow a 7½-inch rise (including the floor

FIGURE 125. *A carriage for a staircase; three are generally required for a do-it-yourself staircase.*

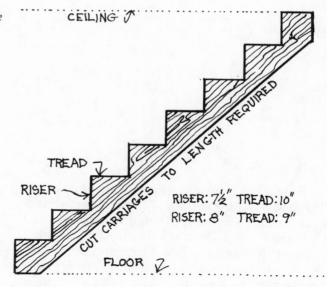

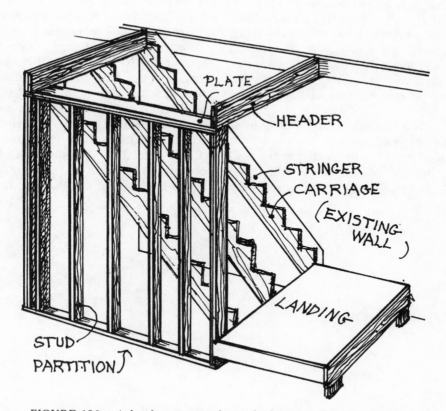

FIGURE 126. *A landing is simply a platform on short posts, commonly in a corner and used to make the staircase turn the corner. Make sure the carriage sits so that its top is at least the height of a riser below the second floor, because this creates the last (highest) step.*

Ups and Downs

thickness). Minor adjustments can be made in the position of the carriage. Then nail one carriage onto any adjacent wall, through wall material and into the studs. If there is to be a wall on the other side of the staircase, build that and secure the other carriage to it. If one side is open, install studs from the floor upward for the outer carriage to rest on, and finish off with plasterboard and wall covering.

Incidentally, while an open staircase is handsome and stylish, it is a heat-waster. It would be better to put the stairs between two walls and put a door at the foot, to prevent warm air from rising to the second floor. This is especially important if the second floor is cooler than the first. Or you can build an open staircase (at least open on one side) and put a door at the top that opens into the second floor.

A staircase with a landing is best — or at least easiest — when put in a corner. Corner or not, the landing is framed with 2 x 8s cleated to the wall studs (see Figure 126). Where there are no wall studs to nail to, the 2 x 8 frame is supported by 2 x 4 posts. If the landing is more than two steps high, it should be supported by 4 x 4 posts or two 2 x 4s clinched together. A landing must be at least 2 feet 6 inches deep to allow for furniture that must be turned around and for an inswinging door if one is put on the landing.

Remember that staircases don't have to be traditional. They can be modern, with or without risers; then can cantilever from the wall. They can be spiral — these are best bought prebuilt. If the spiral isn't very large (6 feet or less in diameter), there should be a straight stair somewhere in the house or outside to permit furniture moving.

For an addition a step or two lower than the floor of the existing house, it is a simple matter to install steps to it; the most important thing to remember is to have sufficient headroom in the entry from the house so that some poor devil walking down those steps does not crack his skull on the overhead. It is even more important to make sure there is enough headroom in the opening to an addition that is one or more steps above the floor of the existing house.

chapter 10

Lights Fantastic

Windows and Doors

WINDOWS

Installation of modern windows in new construction is easy. They are called setup windows, because they come all set up, complete with jambs, sashes, and casing. They come single- or double-glazed (two layers of glass), and in some cases triple-glazed, which makes them resist the passage of heat very well.

Setup windows are ready to insert into the rough openings of a wall. The best type to use is a wooden one, usually primed on the outside and ready to stain or paint. The window is made of pressure-treated wood, resistant to decay from moisture. Some windows are covered on the outside with a thin layer of vinyl, which is good because it makes them virtually maintenance-free on the outside.

Avoid windows of any material except wood. Aluminum and other materials are sturdy enough and fine for storm windows, but as primary windows they may warp, and unless they have a thermal barrier or break (a layer of urethane or other ultra-insulating material between inner and outer sections) to prevent sweating, they will not be good insulators.

There are various types and styles of windows available: double-hung that slide up and down (Figure 127); outswinging casement and awning types; inswinging hopper type; and bay or bow windows (Figure 128). The awning windows are good for a ribbon effect or clerestory arrangement. Hopper windows work well in basements. Outswinging windows are good for directing breezes from outside to inside. There are also fixed windows, which don't do anything except let in light. Bays and picture windows are usually fixed.

There are advantages and disadvantages to the various kinds of windows; in general, a swinging window, one that opens like a door, is easier to seal and weatherstrip than a double-hung window.

Then there are various styles, usually marked by the casing (frame) and number of lights (panes) in each sash. The casing can be so-called Colonial, with ¾ by 3¾-inch pine, or Western, with 1 5/16 by 2-inch pine. These can be used in almost any style of house. So-called picture windows became popular after World War II, but too often they presented a picture for someone looking in from outside rather than for someone looking out. Large expanses of glass today are mainly for solar heat gain. Finally, there are such things as sliding glass doors, but as we have seen these are impractical in a superinsulated house or addition. They are giving way to French doors.

Setup windows are installed before the siding is put on and after the sheathing is. The only

Lights Fantastic

FIGURE 127. *A double-hung window, set up and ready to be put into the rough opening. It can be single-, double-, or triple-glazed.*

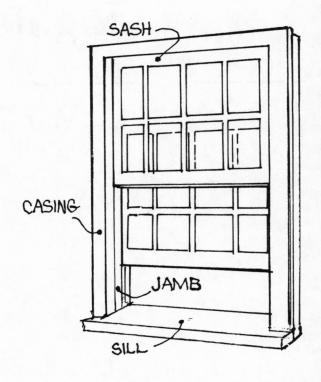

exception to putting in windows before installing siding is when vertical siding is put on over sheathing or horizontal furring strips. The rough opening should be ½ inch wider and higher than the outside measurement of the window jamb, to allow for easy insertion. Before the window is put into the opening, the sides of the opening must be fitted with 8- to 12-inch-wide strips of roofing felt, to assure wind- and watertightness between the outside casing and the sheathing. Leave 12 inches or so of the roofing felt hanging over the bottom of the opening; siding goes over this (described in Chapter 13). When the window is placed in the opening, the casing will press

FIGURE 128. *Top: Outswinging casement and awning windows; inswinging hopper window. Bottom: a bay window.*

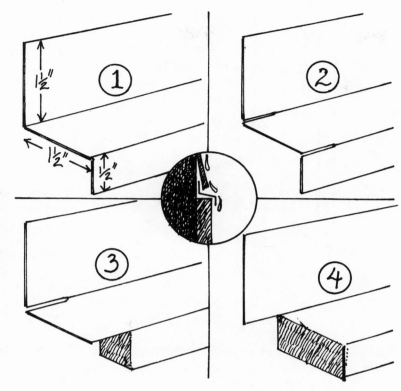

FIGURE 129. *A metal drip cap fits over the top casing of a window, with the top flange under the siding and the bottom flange fitting over the casing itself. It can be cut about half an inch long on both ends, then cut and folded down and back for extra protection to the casing.*

against the roofing felt and against the sheathing, thus sealing any gap. Some setup windows have a gasket that seals this joint. If there is none, you can apply a bead of caulk on the roofing felt so that the casing is sealed to it.

Place the setup window in the opening so that the top sash is flush with the outside of the wall. Some windows have extensions of the sill sticking out beyond the widest part of the casing. These "horns" are to prevent the sill from being banged around in shipment and handling and should be cut off flush with the casing before installation.

The window's sill will sit on the doubled sill at the bottom of the rough opening. Put a level on the window's sill to make sure it is level; if it isn't, shim on the low side. Use low-grade shingles for shims; they are tapered and work well. If the gap that must be shimmed is more than ¼ inch (the maximum thickness of a shingle), use two shingles, one in each direction. When they are driven in from opposite sides, they will bring the window to the proper level. Now check for plumb (vertical), again using a level. Shim if necessary. Nail through the casing into the stud on one side, about 6 inches from the bottom of the side casing. Do not drive the nail home in case you have to remove it. Then check the other side for plumb; adjust if necessary. When everything is plumb and level, nail along each side casing. If the window is more than 2 feet wide, nail into the top casing as well. Nail every 12 inches. Countersink the nails and fill with putty. There is no need to nail through the sill.

Vinyl-clad windows have a special rigid vinyl flange extending from the wood casing. These windows should be nailed with 1½-inch galvanized roofing nails every 4 inches. Predrilled holes in the vinyl flange are a guide for nailing.

Setup windows have a vinyl or aluminum drip cap to guide water down the siding and beyond the top of the casing. Adjust the drip cap so that its horizontal surface is slanting slightly away from the wall sheathing (Figure 129). Nail through the upper flange into the sheathing, using galvanized roofing nails. To protect the sides of the casing, make sure that the drip edge is about 2 inches wider than the casing. Snip into the ends of the drip edge where it folds and fold down and over.

Since the addition is superinsulated, with super-thick walls, there is a problem with windows. Normally the jambs of a setup window are

Lights Fantastic

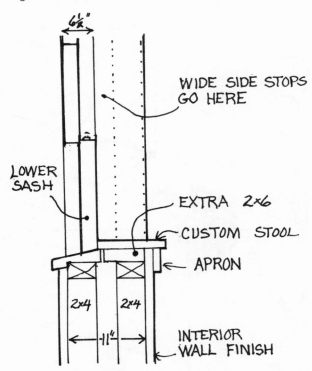

FIGURE 130. *Interior jambs fit in the rough opening of a doubled wall. Extra are the 2 x 6 rough sill and a custom stool, as well as a wide side stop, wide top stop, and the jambs.*

wide enough to fit an ordinary wall. A wall built with regular 2 x 4 studs, ½-inch sheathing, and ½-inch plasterboard on the inside is 4½ inches thick, so the jamb is 4½ inches wide. For a wall built with 2 x 6s, the jamb would be 6½ inches wide.

But our superinsulated wall is 12⅛ inches thick (Chapter 6). To make the jambs fit into this superthick wall, they would have to be 12⅛ inches wide. Impractical, because it is almost impossible to buy pine more than 11¼ inches wide. What to do?

Never fear! Where there's superinsulation, there's a way.

Suppose you buy a window with 6½-inch jambs. This means that the sill will also be 6½ inches wide. When this window is inserted in the opening with a 1-inch overhang on the outside, there will be 5⅝ inches of space on the inside at the bottom (Figure 130). This opening can be filled with a 2 x 6, which would bring the sill to the edge of the interior wall. A trim of nominal 1-inch board (called a custom stool) can be put right on top of this 2 x 6, so that it sticks out 1¾ inches beyond the inner wall, allowing an apron (called a stool cap) to go under it.

How wide should the 1-inch board be? The wall is 12⅛ inches thick. Add 1¾ inches overhang to bring the board to 13⅞ inches. But of course it doesn't go under the sashes; it butts against the bottom sash. So subtract the thickness of the two sashes (3¼ inches), and you get 10⅝ inches. If you use a casement window, with only one sash thickness, the width of the board is 12¼ inches. Whatever the width, it is best to measure it when everything is in place. It will probably have to be ripped (sawed lengthwise) to a custom width.

The side stops are a little easier. They are 8⅞ inches wide—obviously a custom cut. The top stop is 10½ inches wide.

These far-out dimensions add up to nothing more than very deep window openings. That space can be put to good use: the least expensive, practical one is the installation of inside storm windows, set up on the interior casing (Figure

FIGURE 131. *An inside storm window mounts on the inside casing the way an old-fashioned wood storm mounts on the outside casing. It is tightened with spring-loaded clips.*

 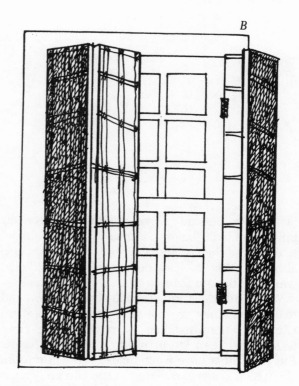

FIGURE 132. *Thermal shutters can be covered with wallpaper or other decorative things to make them noticeable; or finished plainly, to make them obscure (A). They also can be bifold (B).*

131). You could get fancy and build a stop inside the side pieces and mount the storm window there.

A good inside storm is made of glass with an aluminum frame. It has rubber gaskets and hangs from the top casing like an old-fashioned wooden storm. It is held against the casing with spring-loaded clips. Another type is magnetic, made of acrylic with a plastic or metal frame, held onto the casing with a magnetic strip. These are generally good when there is no inside wooden casing. But their disadvantages are that they must be custom-installed, and that the frames must touch the casing on all four sides and along the entire length of each side, in order for the magnets to work. Also, the acrylic surface is sensitive to scratching and other blemishes, particularly if these windows are stored in a basement or garage over the summer.

Inside and outside storm windows with a double-glazed main window will give each window quadruple (quad) glazing, highly resistant to heat loss for a see-through window. Another way to get the same result is to install a double-glazed window on the inside of the window opening. The only problems are that such windows haven't been designed yet; and they would double the price of each window. Regular setup windows would work, with the outside casing set up on the inside, the bottom sill replaced with a nonslanting 2-inch board, and side and top stops for it to butt up against. But it still doubles the price.

Ideas, however, like hope, spring eternal. Inside thermal shutters can be placed in these deep openings. They can be made quite decorative, or can be unobtrusive, possibly even self-storing, by folding out along each side casing (Figure 132 A) or by having two folds so that they don't stick out beyond the inside casing (Figure 132 B). (See Chapter 3 for more about thermal shutters.) People who have visited European homes with thick stone walls or pre-Colonial houses in this country with similarly thick walls will not find the extra-wide jambs and the shutter treatment unfamiliar.

Whether you install double- or single-glazed windows, or use inside storms, adding permanent aluminum storm windows on the outside is a good

Lights Fantastic

idea. New outside storm windows reduce air infiltration to far below national standards, almost to zero. And they protect the regular window sash, although this is less necessary for vinyl-clad windows than for painted or stained wooden ones.

While outside storms reduce air infiltration and exfiltration, they are not fully airtight. They are caulked when installed to prevent rain from leaking between storm frame and wood frame. But they must have weep holes in the aluminum sill, which is a separate unit at the bottom of each window. Weep holes allow rainwater that does get between storm and main window (during the summer, for example, when the bottom storm sash is up and a screen replaces it) to drain away. The weep holes also allow the window to "breathe." Here is how it works: when air between storm and main window heats up (as on a sunny winter day), it expands. When it cools (when the sun goes down or is obscured by clouds), it contracts. Expanding air needs somewhere to go, and the weep holes are relief valves, preventing that expanding air from popping something. Even more critical is that cooling air may contract enough to create a partial vacuum, enough to break the glass if there are no relief valves. This may sound far-fetched, but breakage can occur. Of course, it can happen only if both the storm window and the main window are virtually airtight.

DOORS

An exterior door comes set up the way windows do, complete with door, jamb, casing, trim, hinges, and locks (Figure 133). It can even include sidelights and fancy trim for outside finish. Doors come in fir, pine, and insulated steel.

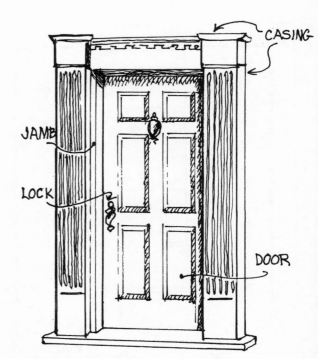

FIGURE 133. *The anatomy of an exterior door.*

If your house is not very old or historical, you can install an insulated steel door.

Standard door heights are 6 feet 8 inches and 6 feet 6 inches. Front doors are 32 to 36 inches wide and service doors 30 to 32 inches wide. Avoid double doors (two doors with no vertical post in the middle). They are a security risk. You must specify the door to fit the opening.

The sill, or threshold, is 2 inches thick and made of oak. Today's sills are designed to sit flat on the subfloor, with the top side sloped to shed water. Door jambs are cut to follow the contours of the sill. The jambs are made of 1⅛-inch-thick wood, or thicker, with a door stop built right in (Figure 134). This integral stop promotes security and weathertightness. The jambs are really 2 x 6s (5½ inches wide). More expensive doors are weatherstripped with integral, interlocking flanges that really keep out the weather. Otherwise, weatherstripping must be installed on the outside of the door. Steel doors have their own magnetic or gasketed weatherstripping.

Doors are installed so that they are indented about 3½ inches in from the outer wall, even if the wall is doubled and 12 or so inches thick. With an extra-thick wall, however, there must be ex-

Windows and Doors

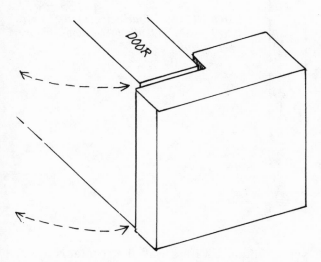

FIGURE 134. *The stop of an exterior door is built into the jamb.*

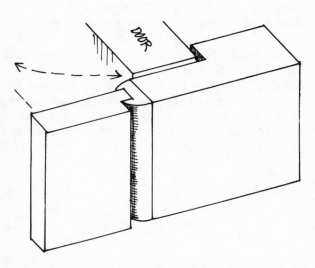

FIGURE 135. *Jamb extenders allow a door to fit into an extra-thick wall.*

tensions on the jambs, just as there are extensions on the window jambs. The sill need not be any wider, because of the location of the door itself, and the floor and finish-floor material simply are fitted up to the indentation of the door opening.

Door jambs are available with extenders (Figure 135) that fill the gap created by the door's position in an extra-thick wall. The extenders may still not be wide enough for the super-thick wall, so filler pieces can be attached and the joints filled with putty.

Since a door is hinged on the inside and swings in, the extra-thick wall will prevent it from swinging completely open; that is, it will not swing 180 degrees to set against the interior wall. It may swing a little more than 90 degrees, but not much. This is okay for most purposes. One way to get around it, if necessary, is to reduce the wall on each side of the door to normal thickness and to install sidelights in the thinner wall portion (Figure 136). The loss of insulation will be minor. If the door is in an entryway that is as wide as the door plus sidelights, the thinner wall will not even be noticed. And if the entryway has an inside door, making the area an airlock, it will be snugger than a superinsulated wall.

Whatever the door position, the technique for installing it is similar to that for windows. Put roofing felt strips on each side of the opening. Nail through the side jambs into the studs of the rough opening. Make sure the threshold is level and the sides are plumb. Also make sure the door is square by placing a carpenter's square in each upper corner of the opening. With everything square, plumb, and level, shim with shingles where necessary and complete the nailing. Pairs of nails should be driven in every 16 inches along the jambs. Add the outside casing after everything

Lights Fantastic

FIGURE 136. *Sidelights set in an extra-thick wall will allow the door to open nearly fully.*

else is installed. The inside casing goes on after the finish floor is laid down (see Chapter 17).

Steel doors go up the same way, except that the frame which goes in the rough opening is made of steel. This frame should be installed in a wall of standard thickness. A wooden casing can be put over and around the steel frame. Steel-insulated doors do not need storm doors; in fact, they shouldn't have them because on sunny days the air can get so hot it can cause paint to fail. A wood door should have a weatherstripped storm door. Wood is better than aluminum, but it has to be painted and *kept* painted. A steel door can be equipped with a screen door in the summer.

One more point about windows and doors: if your addition covers a wall that contains a window, remove it, frame and all, and install studs to fill the opening. Cover with plasterboard to match the wall finish inside the existing house and also in the addition. Cover the edges of the plasterboard patches with joint compound and paper tape. With practice you can make an invisible patch.

You might be able to use the window you removed for one in the addition. Or, just for kicks, you might want to leave the window in place to allow light in the existing house or as a decorative touch. Still another choice is to remove the window and rebuild the opening for a door or doorway into the addition.

chapter 11

Icing on the Cake

Outside Trim

Exterior trim is just what it says: the finishing touch to the outside of the addition. It is more than aesthetic, however; it covers various parts of the building that need covering, to prevent weather from getting in and causing decay.

Trim includes the eave assembly, or cornice, where the roof overhangs the wall; the rake (gable) end of the roof, where the edge of the roof meets the gable wall; and cornerboards. Even if trim is the finishing touch, it is put on before siding. In this chapter we discuss eave and rake trim; cornerboards are discussed in Chapter 13.

Trim is usually made of clear pine; if there are knots in the wood, it is necessary to seal them with shellac before painting or staining. This is a time-consuming chore and there is no guarantee that knots won't bleed through the paint or stain anyway. Thus, the more expensive clear pine is the wood to use. If possible, soak the ends of trim boards in a wood preservative before installing. Ideally, the inside of the trim, that is, the sides and edges that will not show, should be painted with an oil-based exterior primer before installation. The more you do this, the longer the trim will last. But it should last indefinitely even if you don't, so long as it is well painted or stained where it is exposed to the weather, and is kept dry where it is not exposed.

Generally, use galvanized box nails for nailing; their heads can be countersunk and the holes filled with putty. Casing nails are better; they have heads smaller than box nails but larger than finishing nails. They may, however, be difficult or impossible to find.

The most obvious trim on the addition is at the eaves. This is the cornice, the extension of the roof and the (usually boxed) structure that connect the roofline to the wide walls (Figure 137). The face of the cornice is called a facia, and the underside is called a soffit, which is often supported by a frieze board.

The width of the cornice is determined by the extension of the rafters and their pitch (see Figure 92). The shallower the pitch, the wider the overhang can be. If the pitch is steep, then the soffit of the overhang might extend below the top edge of the windows.

The overhang on a wall that is mostly glass, for passive solar heat (usually the south side), should be 30 inches (see Chapter 2), to block the summer sun when it is high in the sky and to let in the winter sun, when it is low. The overhangs in the rest of the addition can match the existing house or can be narrow or wide as you choose.

If the rafters are cut flush with the wall (Figure 92 A), the cornice is called a close cornice. Although it is easy to build, it is skimpy, provides no overhang at all, and also, because there is no

Icing on the Cake

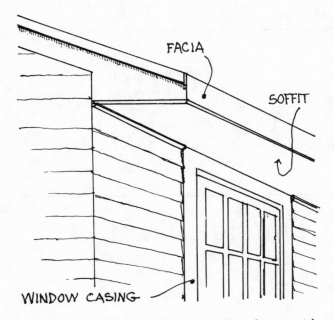

FIGURE 137. *A cornice is simply a boxed eave, with rafters overhanging the wall, faced with a facia board and bottomed out with a soffit.*

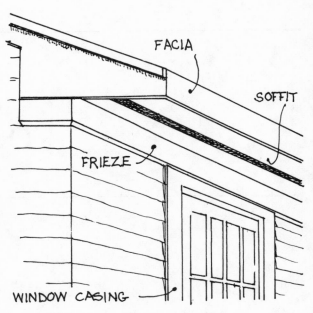

FIGURE 138. *A 2-inch ventilation strip is made by screening along the bottoms of the rafters and applying the soffit in two pieces.*

soffit, is difficult if not impossible to ventilate. It is not recommended.

You can build a relatively narrow cornice or overhang, from 6 to 12 inches (Figure 92 E), or slightly wider (Figure 92 F). This type of cornice requires rafters cut both plumb and level, so that the plumb cut can be covered with the facia and the bottom, level cut with a soffit. The facia usually overhangs the soffit by ½ inch.

The soffit must be ventilated, especially in a hip roof. Screen the bottom of the rafters with ordinary fly screening. Cut the soffit in two pieces and install them so as to leave an opening 2 or more inches wide down the middle of the entire length of the soffit (Figure 138). You can use nominal 1-inch lumber or ⅜-inch plywood. Do not use ¼-inch plywood. It will bend and waver and look terrible. When nailing finish plywood, use siding or shingle nails. It is difficult to countersink nails in plywood, and this plywood is too thin to countersink nails into, anyway. Shingle or siding nails can be nailed flush and painted or stained and will not show.

Cornices 6 to 12 inches wide are enough to protect the side walls and to give the roof a substantial look. A wider cornice is usually more contemporary in design. It is built in the same way as a narrow one, except that the soffit is nailed not only to the rafter bottoms but also to a 2 x 4 or 1 x 4 ledger (Figure 92 E). In this case the ventilation strip cannot be in the middle of the soffit. You should therefore position the strip at the front of the soffit, just inside the facia.

For an extra-wide overhang, there is still another technique (Figure 92 F). Lookout members of 2 x 3s or 2 x 4s are nailed to the side of each rafter end and then toenailed into the sheathing, preferably into a stud. A 2 x 4 ledger will help secure the lookouts. Then the soffit, made of two pieces to allow the ventilation strip, is nailed to the bottom of the lookouts.

Next comes the facia, a nominal 1-inch board, which is nailed to the rafter ends (Figure 139) and must be wide enough so that it extends ½ inch below the soffit. When positioning a facia board, hold a block of wood along the roof and use this to align the top of the facia with the roofline (Figure 140). A 1 x 2 is nailed at the top of the facia (Figure 141), to act as a drip edge and extend the roofline.

The final piece of the cornice is the frieze board. It goes against the sheathing just under the soffit (Figure 142). If the windows are positioned properly, there will be enough room for the frieze board to go over the top casings.

When you get to the rake of the roof, where it

Outside Trim

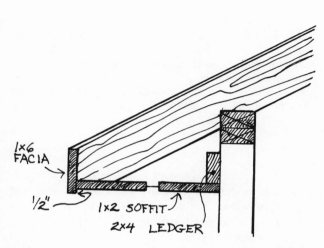

FIGURE 139. *The facia fits at the ends of the rafters; soffit sections are nailed to the bottom of the rafter and to a 2 x 4 ledger.*

FIGURE 140. *When nailing on the facia, place a scrap board on the roof so that the facia will be positioned right, following the angle of the roof.*

FIGURE 141. *To increase the dripability of the drip edge, add a 1 x 2 at the top of the facia.*

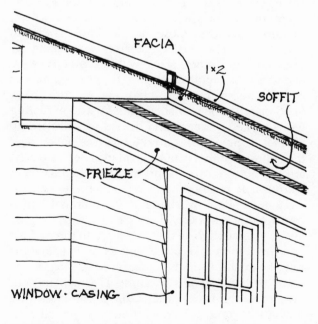

FIGURE 142. *Sometimes, according to the location of the tops of the window casings, a frieze board can be placed under the soffit, sitting on top of the casings and finishing off the house trim.*

Icing on the Cake

meets the gable wall, things get a little more complicated. You can have a close rake or an overhanging one. If you look around your neighborhood, you will notice that houses built more than thirty years ago have rake overhangs. More recent ones have close rakes, because they are cheaper to build and use less material. In the good old days of cheap material and low labor costs, overhanging rakes were a matter of style. It's best, therefore, to have a close rake, and here is how to make one that is, or that just slightly overhangs. A 1 x 6 board (other sizes will do, and may be necessary to achieve the right proportions) is nailed directly on the sheathing (Figure 143), lined up with the roof sheathing. A strip of roofing felt is tucked under it, and the siding is butted up against it. Sometimes a rake board is nailed over the siding, but this is not a good idea because gaps are left between the board and the clapboards or shingles. It is sloppy and not recommended.

To get a slight overhang, nail a nominal 2-inch board along the rake, then a trim board (Figure 144), wide enough so that it is flush to the roofline and the 2-inch rake board and extends below the rake board by ½ inch. Make sure all boards are flush with the roof sheathing. If you plan your rake properly, the roof sheathing will extend far enough beyond the end rafter to cover all rake pieces. Because the 2-inch board is solid, there is no need for a soffit board. To bring the rake out another 1½ inches, nail short nominal 2-inch blocks to the first board. To bring it out 3 inches,

FIGURE 143. *A close rake is a trim at the end of the roofline at the gable; it is a 1 x 6 or larger board nailed directly on the sheathing, positioned at the edge of the roof sheathing. With a little foresight, you can extend the roof sheathing over the edge so that the rake trim can be butted up against it.*

just add another nominal 2-inch board, and add the trim board. Because of the short pieces, you will need a nominal 1-inch soffit board. Quarter-inch plywood will do here.

Where the trim board meets the eave cornice, you have to make a neat corner. This is called an eave return, and in the case of a close rake is simply an added piece of trim to make a tidy appearance (Figure 145). It can be curved or cut at an angle.

For a greater overhang, 6 to 10 inches, you have to plan ahead and extend the roof sheathing beyond the end rafters (Figure 146). Then toenail short 2 x 4 or 2 x 6 lookout blocks through the wall sheathing into the end rafters, on 16-inch centers. Their ends are connected by a facia board. This, plus the attached overhanging roof sheathing and the addition of a soffit and a frieze board, will

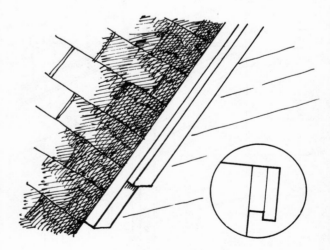

FIGURE 144. *A moderate rake overhang is achieved by nailing a nominal 2-inch board along the rake, and covering it with a nominal 1-inch trim board. A doubled 2-inch board will bring the rake overhang out another 1½ inches.*

Outside Trim

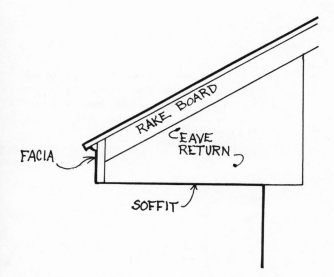

FIGURE 145. *The anatomy of an eave return.*

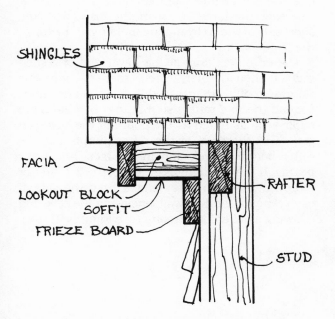

FIGURE 146. *A greater overhang is achieved by extending the roof sheathing first. Then lookout blocks are attached to the end rafters and covered with a facia and soffit.*

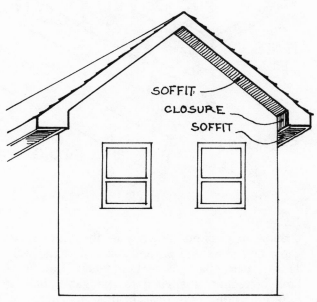

FIGURE 147. *To accommodate a rake overhang, the eave return is boxed.*

make the lookouts stable. For added strength, particularly when the lookouts are longer than 10 inches, nailing blocks are set between them and nailed to them. Also, a fly rafter of nominal 2-inch lumber is nailed to the lookout ends and to the ridgeboard, which is extended just as the roof sheathing is.

An eave return with an overhanging rake is more complicated than the one for a close rake. The soffit of the eave is extended beyond the corner to meet the roof edge. Then to protect it from roosting pigeons, it is roofed and boxed in, with the boxing material reaching to the house side and the soffit of the overhanging rake (Figure 147).

Icing on the Cake

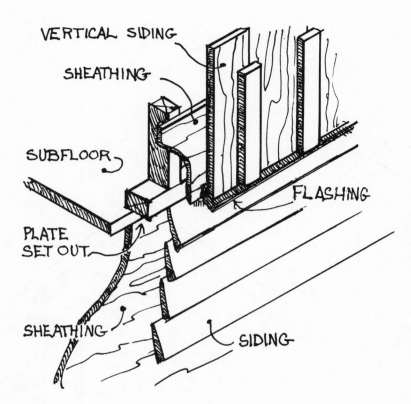

FIGURE 148. *To change the kind of siding at a gable wall, extend the upper section out slightly. Flashing is tucked under the upper section and over the lower section.*

Some styles of building require the entire upper gable end to extend beyond the rest of the wall. This is particularly important when a different kind of siding is applied to the gable wall. A different siding (vertical boards, for instance, when the rest of the siding is clapboard or shingle) is used when a gable is large, for simple style reasons. The easiest way to extend a gable wall is to nail 2 x 2s directly on the gable studs, then apply the sheathing. Or nail horizontal 2 x 3s or 2 x 4s to the gable studs, on which to nail a vertical siding.

If you like the idea of changing siding at the gable wall and can't or don't want to make the upper wall extend over the lower wall, it can be done this way (Figure 148): install metal flashing in an L shape, with the longer side of the L extending over the upper siding and the narrow side over the lower. Support the drip edge (the short side of the L) by nailing molding at the top of the lower siding.

The eave overhang at the lower end of a shed roof or shed dormer, where there is only one slanting roof, is, thank goodness, a simple matter. Extend the roof rafters so that they overhang the outer wall, as much as the depth of the eave you plan. Cut the ends of the rafter plumb, cut the rafter bottoms level, and install facia and soffit boards. If the overhang is extra deep (12 inches or more), 2 x 4s can be extended from the side of each rafter to the side wall, enclosing the eave, and allowing a nailer for the soffit (see Figures 91, 92).

chapter 12

Top o' the Mornin'

Roofing

When you have the frame of the addition up and the sheathing on the roof and walls, it is a good idea to put it "to the weather" as soon as possible; that is, protect it from the elements by roofing and siding. Roofing is more critical, so we'll tackle that first.

There are basically only two types of roofing: multiple and membrane. Multiple roofing is shingles, used on sloped roofs, such as 4-in-12, or 4 inches of rise for each 12 inches of run, or steeper. The shingles protect the building by shedding water. Most are made of asphalt, although wood shingles or the big wooden "shakes" claim a small share.

Membrane roofs are for flat or nearly flat roofs, and are built up or made of roll roofing. The membrane on a dead-flat roof must be thoroughly waterproof because it holds the water.

You'll want to match the type of roofing on your addition to that on the main house.

Asphalt shingles are the most commonly used and will last twenty years or more with virtually no maintenance. Wood will also last for that length of time with very low maintenance. Asphalt shingles come in various shapes and weights. The most common weight is 235 pounds, meaning the shingles for each 100 square feet (called a square) installed weigh 235 pounds. They are made of asphalt-impregnated felt covered with a layer of asphalt onto which mineral granules are applied for resistance to wear and tear from weather. There are shingles of heavier weight (240 pounds and as high as 300 pounds), some of which are quite fancy. The heavier-weight shingles will last longer than the lighter-weight, but consider them for how they look only if your addition or house has a large expanse of roof readily seen from the ground. You wouldn't want to put a fancy roof on a three-story-high hip roof. Only God and airline pilots will see it.

All asphalt shingles are self-sealing. They have a strip of asphalt on the back. When the shingle is installed, the warmth from the sun melts the asphalt, which acts as an adhesive. Such shingles are generally resistant to being lifted by a strong wind.

Color is a matter of taste, although you may want to match the color of the roof of the existing house. But if you plan to reroof the house, this might be the time for a change. A lighter roof will reflect a certain amount of heat, but this is significant only in southern climes. It is virtually insignificant if the attic floor is insulated and properly ventilated, which, in a superinsulated addition, it is likely to be.

A light roof will make a building look taller than

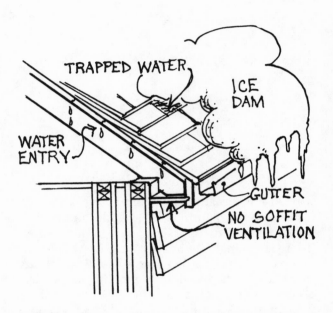

FIGURE 149. *Ice dams are often caused when there is no insulation and no attic ventilation.*

it is, while a dark or black roof will keep it from soaring too high to the eye, and will make it look larger.

A common plague in northern climates is ice dams, usually in areas of moderate to heavy snow in the winter, and particularly troublesome when weather is alternately warm and cold.

Here is what happens in an ice dam (Figure 149). Snow builds up on the edge of the roof at the eaves. The more snow, the heavier the buildup gets. If it begins to melt from the bottom up, perhaps because of a warm attic, the melting snow turns into ice during cold spells, forming a dam along the eaves. When snow continues to melt, the water is stopped in its downward flow by the dam, and can back up, uproof, until it finds its way under the shingles and through the sheathing, into the attic, onto the insulation, and into the house. It could also find its way through cracks in the boxed cornice where the roof meets the wall.

These dams are usually caused by a combination of the roofing and the design of the house, so the solution, like the problem, involves them both (Figure 150). First the design: a well-insulated attic and good ventilation, particularly along the soffit, will keep the attic cold, preventing melting of snow from the bottom.

As for the roofing: a 36-inch strip of roll roofing or 45-pound felt along the eaves, for the entire width of the roof, will prevent the ice from backing up and meltwater from going under the

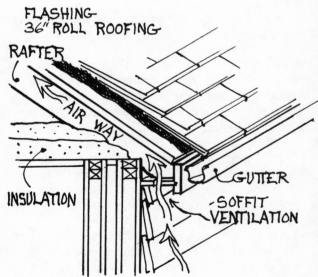

FIGURE 150. *With insulation, soffit ventilation, and 36-inch roll roofing at the eaves, under the shingles, ice dams are kept to a minimum; even when they occur, the roll roofing keeps water that backs uproof from getting under the shingles and into attic and house.*

FIGURE 151. *An aluminum drip edge is good along the rake of a roof as well as along the eaves.*

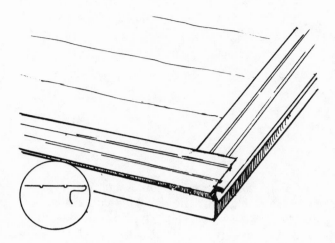

shingles. The shingles will actually go over this strip, and the water could get under them, but it will be prevented from penetrating the sheathing by this roll-roofing strip. And the chances are pretty good that water will not be forced uproof for more than the 36-inch width of the strip.

Before you start laying shingles, there are important things to consider: the edges of the roof and what goes on underneath the shingles.

Edges require a drip edge, a folded piece of metal that fits along the eaves and the rake edge (Figure 151). The best one to use is aluminum. It does just what its name says: allows an edge over which water can drip. Its function is improved if you have nailed a 1 x 2 trim board along the eave facia and rake board. The drip edge takes the place of wood shingles placed at the eaves, a past practice. Wood shingles tended to rot when put under asphalt ones.

It is best to use aluminum nails when nailing an aluminum drip edge. If you use galvanized nails, the two metals in contact with each other set up a galvanic action that could hasten corrosion. Some roofers pooh-pooh this, however, admitting that while it may be true, the corrosion will take longer than the roof will endure.

An aluminum drip edge should be installed under roofing felt along the eaves and on top of the felt at the rakes.

What about roofing felt? Fifteen-pound felt is sometimes laid on the entire roof over the sheathing, either single (overlapped by 2 inches) or double (overlapped by 19 inches). The double coverage is most important on shallow roofs. But roofing felt is not always necessary under asphalt shingles when the roof is fairly steep, and is certainly not needed over plywood sheathing on really steep roofs. It may, however, be required by a local building code. One of the reasons some roofers don't like to use felt is that it reduces the ability of the roof to "breathe," and if moisture is trapped under it or the roofing, shingles will tend to curl. These curls are called "smiles," but neither the householder nor the roofer is smiling.

Now for the shingles. Normally, they are applied directly to the roof sheathing, but when insulation is put on the top of the roof, over the sheathing (as in the case of a cathedral-ceiling roof with exposed beams), the shingles are put on top of the insulation, using long nails (see Chapter 3).

A common asphalt shingle is 12 inches wide and 36 inches long, with two slots nearly half the depth of the shingle (Figure 152). Each end has

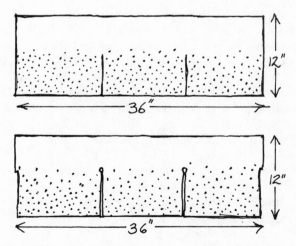

FIGURE 152. *Asphalt shingles: (bottom) with slots; (top) with score marks instead of slots, an improvement because shingles erode and weather first in the slots. Newer shingles are fiberglass, slightly larger than asphalt shingles.*

half a slot so that when it butts against the half-slot of its neighbor, the divisions come together and look like a whole slot. The slots are designed to make the shingles look like wood or slate shingles.

Another common asphalt shingle has no slots but does have shallow score marks. They are less distinct than the slots and tend to wear better than the slotted ones do; in the past it was discovered that when early wear occurred, most of it was in the slots themselves.

Still another type is the fiberglass shingle, with a fiberglass core instead of saturated felt. This shingle is larger than the asphalt type, and therefore fewer are needed for the same coverage, so roofing jobs can be done faster. They are generally designed to last somewhat longer than asphalt shingles, even though both types are covered with asphalt and the mineral granules. They are likely to replace asphalt shingles.

Either type is easy to apply. Fancier, heavier-weight shingles will wear better and look quite different from common shingles, but they all go up the same way. They butt against each other and are nailed by a certain number of nails. Common shingles are designed to overlap by 7 inches, leaving 5 inches of exposure. This will give a double coverage; that is, there are two layers of shingle on each course (row). If you overlap the shingles by 8 inches, leaving 4 inches exposed, you will get triple coverage: three layers of shingle on each course.

Install all shingles at edges so that they overhang the drip edges along the eaves and rake by ¼ inch. The shingles will not droop, even in hot weather, and will not break unless they are struck with a hard object. They can also be installed flush with the drip edge.

Lay the first row of shingles with their top sides toward the eave (Figure 153). This is the under-

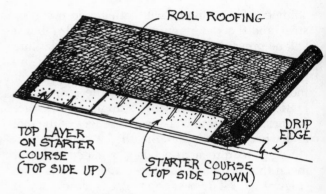

FIGURE 153. *Under the roll roofing at the eaves, and over the drip edge, the starter course of shingles is top side down; first course goes on top of it, top side up.*

course. Then lay the first, or starter, course right side up, directly on top of it. When installing the undercourse, cut the first shingle in half. Then, when the first course is laid directly on top, starting with a full shingle, the joints will not line up. If you laid all the courses beginning with full shingles, you would get all the joints and slots directly over each other—something you obviously want to avoid.

Incidentally, most asphalt shingles come in bundles with excellent installation instructions printed on the label. Most shingles have small slits on each side; when these slits are opened they act as little hooks so that the next course can be automatically aligned. The slits are set for a 5-inch exposure, the most economical and most common shingle installation.

Use galvanized roofing nails 1 to 1¼ inches long. Drive four nails into each shingle, one above each slot and half-slot, about ¾ inch above the slot. The same number of nails is used when

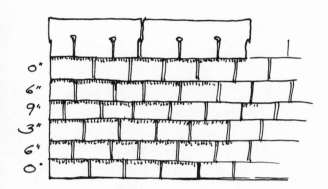

FIGURE 154. *The order of cutting shingles is important to make sure the slots of each course line up only every sixth course.*

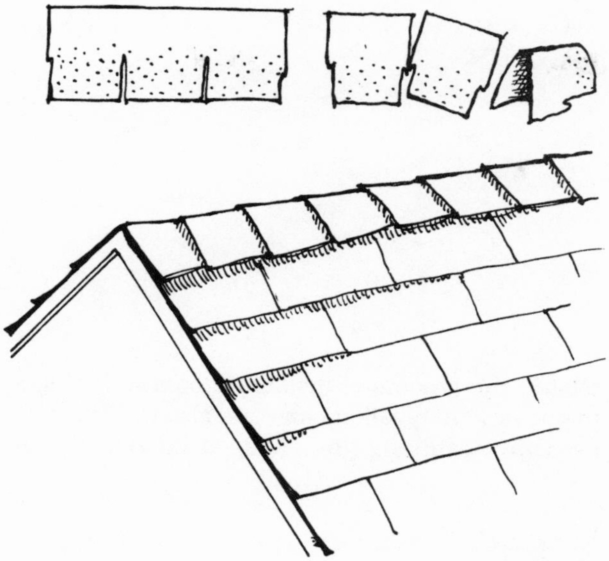

FIGURE 155. *Ridge shingles, 12-by-12-inch squares cut from regular shingles and folded over the ridge, half-lapped, make a Boston ridge.*

installing the slotless shingles, in the same approximate position. The nailheads will be covered by the next course.

A good roof has slots staggered considerably, so that they line up only on every sixth course (Figure 154). To do this, don't cut the first shingle in the starter course at all; cut the second course 6 inches; the third course, 3 inches; the fourth course, 9 inches; the fifth course, 6 inches; and the sixth course, 0 inches. Repeat this cutting pattern as you progress.

Cut shingles with a utility knife, and use a square to make straight, square cuts. Cut on the back side; the gravel on the top side will dull the knife quickly. You can sometimes score the shingle with the knife, then snap or break it.

The slits on each side of the shingles, as we said, will allow you to line them up at the 5-inch exposure. But measure the exposure every now and then, more often than not, to make sure it is consistent. Of course, *don't* step back to admire your work or even to check it. Get down from the roof and eyeball the work to make sure the exposure is, indeed, proper. If it is not, take off the shingles and do it right. If you goof on one of the lowest courses, the entire roof is going to be out of whack.

Shingles are nailed the same way for each course. When you get to the ridge, the last course may overhang it. In that case let it do so until you shingle up the other side of the roof. Then fold one overhang down over the other.

Next, put on a ridge. It is called a Boston ridge and it works this way (Figure 155): cut shingles into three parts; each will be 12 inches by 12 inches. Fold one over the ridge so that the gravel side is at the end. Nail it, and place a second ridge shingle over the first, leaving a 4- or 5-inch exposure. Do this down the length of the ridge. The last shingle must be face-nailed; that is, with the nails showing. Cover the exposed nails with a dab of roofing cement.

Try to determine the direction of the prevailing winds in your area, and nail the ridge squares so that they are closed to this direction. Hips are done the same way as ridges, except that the open end of each square faces the down slope.

If you plan a ridge vent, you have already left at least 2 inches open (without sheathing) on either side of the ridgeboard. The ridge vent is simply nailed through the flange on each side into the shingles and sheathing just below the opening on each side of the ridgeboard. Some vents do not need roofing cement between the flange and the shingles, but it wouldn't hurt.

Roofing cement, by the way, is going to be used

a lot in the process of installing roofing. It comes in one- and five-gallon cans and also in cartridges used in a caulking gun. Keep it handy.

Flashing

As critical as the roof itself is flashing, used in the valley where two sloping roofs meet, and where the roof is connected to such things as vertical walls and chimneys.

A valley can be treated with either exposed flashing or flashing with shingles interlaced across it (Figure 156). Either way is effective; the exposed valley is neater. Where two roofs of equal pitch meet, roll-roofing flashing is folded so that equal widths extend on each side of the valley. Use an 18-inch strip facedown, with 9 inches on each side, and a 36-inch strip face up on top of it, with 18 inches extended on each side. Shingles are applied to the valley with the exposure of the flashing 4 inches wide at the top (2 inches on each side) and widening toward the bottom at the rate of ⅛ inch per foot. Don't crease the roll-roofing flashing; let it gently span the valley.

You can also use metal flashing. Copper is very expensive but good-looking. Aluminum is your best buy. The total width of the metal flashing should be 12 inches for roof slopes of 7-in-12 or more, 18 inches for 4-in-12 to 7-in-12, and 24 inches for slopes less than 4-in-12.

Where a valley connects one steep roof with one shallow roof, metal flashing should be used with a 1-inch standing seam (Figure 157) in the center. This is to prevent water rushing down the steeper slope from pushing up under the shingles of the adjacent, shallow slope. In all cases, install a ribbon of roofing cement under each edge of shingle.

For a shed roof that butts up against a vertical

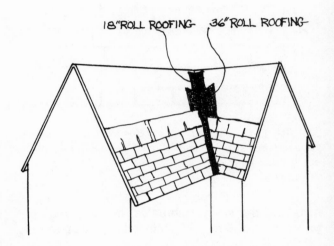

FIGURE 156. *A valley, with flashing under the shingles. A valley can also have the flashing exposed or the shingles laced across it.*

FIGURE 157. *A standing seam must be made of metal. It is used in a valley connecting a shallow sloped roof with a steep sloped roof. The seam prevents water coming down the steep roof from crossing the valley and going under the shingles of the shallow roof.*

Roofing

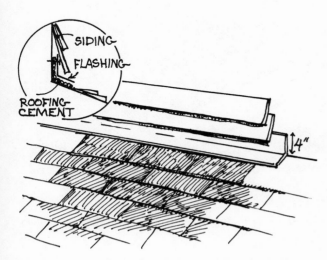

FIGURE 158. *Flashing is easily applied in an L shape where a shed roof butts up against a vertical wall.*

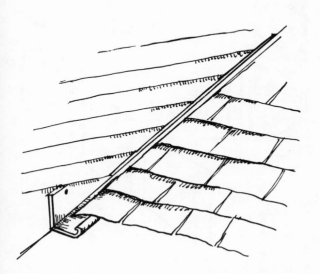

FIGURE 159. *New is flashing with a rolled seam, with roof shingles on top of it (exaggerated for clarity). The water stays in the flashing and finds its way to the drip edge.*

wall, the flashing is easiest (Figure 158). Simply fold a 12-inch-wide aluminum strip into an L shape and nail one leg of the L (4 inches high) against the wall sheathing. Cover it with siding. The other, 8-inch, leg of the L sits on top of the shingles, which come up to the wall. Apply plenty of roofing cement on this seam. If the flashing tends to wrinkle or waver, nail it in strategic places, and cover the nailheads with dabs of roofing cement.

There are two kinds of flashing to use for a sloping roof butting up along a vertical wall: rolled flashing and step flashing. Both are made of aluminum.

Rolled flashing is easier to install (Figure 159). It consists of a length of 12-inch-wide aluminum folded in half and put in the corner where the roof and wall meet, before roofing shingles are attached. Siding goes over the flashing on the wall. The other leg of the flashing has a rolled edge, on top of which the roof shingles are put. The rolled edge acts as a dam, preventing water that gets under the shingles from going through the sheathing. Shingles are butted against the leg of the flashing on the roof and butted up to the flashing portion on the wall, and roofing cement is applied under them. Be careful in your work that you don't bend the rolled edge down, or else it will lose its damming ability. You may have to roll the edge of the flashing yourself. It is simply a matter of folding up the edge about $3/16$ inch.

More reliable, if more complicated, is step flashing (Figure 160). Step flashing is made up of individual pieces of aluminum. They are of different sizes, but 12 by 12 inches is recommended. A piece is folded in an L shape and applied to the joint between roof and wall. The starter course of shingles is nailed over this. The second piece of flashing overlaps the first by 7 inches, and the second course of shingles is then applied. Step

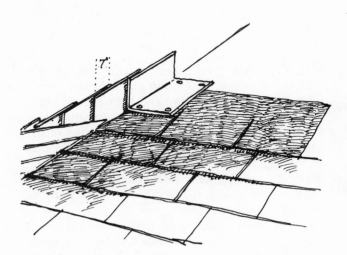

FIGURE 160. Step flashing: one piece goes over each course of shingles when the side of a slanting roof butts up against a vertical wall.

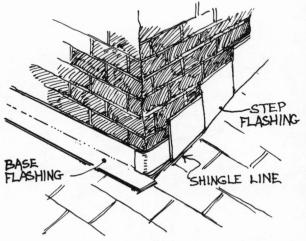

FIGURE 161. Step flashing and base flashing along a chimney.

flashing can be put on after roof shingles are installed as long as nails are not driven too close to the wall. Siding is applied to the wall over the wall flashing, after all of it is installed.

If the abutting wall is brick, or if the roof goes up the side of a chimney, the step flashing is applied in the same manner (Figure 161). Where the roof descends to the high side of the chimney, a full length of flashing is applied, with roof shingles on top of it. On the low side of the chimney a similar strip is used, sitting on top of the roof shingles. Generous amounts of roofing cement are applied under the flashing. Now counterflashing is installed (Figure 162). This is a preformed piece of metal with a lip on its upper side. It is straight for the high and low sides of the chimney, and either cut like a stair carriage or used in pieces for the sides. The lip on the upper side is what makes this flashing weatherproof. The lip is inserted in a joint between bricks and is mortared. The bottom of such flashing stops just above the shingles.

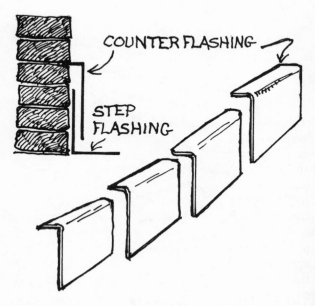

FIGURE 162. Counterflashing covers step flashing; its top edge is bent so that it can be inserted in a brick joint and mortared in place.

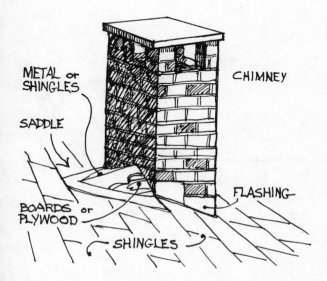

FIGURE 163. *When a chimney is in the middle of a large sloping roof, or when the chimney is especially wide, a saddle will keep rain from piling up against its high side and will help divert snow.*

Sometimes when the chimney is particularly wide or the roof particularly large, a saddle is installed on the high side of the chimney (Figure 163). This diverts rain and prevents a pileup of snow on the high side, which could cause problems. The saddle is made of wood with shingles or roll roofing as a protective covering, and with flashing or a full metal roof over it.

Plumbing vent pipes and roof vents that must pierce the roof come equipped with collars that are their own flashing. These are installed with the collar under the shingles on the high side and over the shingles on the low side.

A skylight or roof window also has its own flashing, again placed under the shingles on the high side and over them on the low side. Flashing along the sides of these installations is generally rolled.

Wood Shingles

Wood-shingled roofs are good-looking and should be used either where matching the roof of the main building is important or where the roof is easily seen from the property around the house or addition. Otherwise the good looks will be wasted. Check your building code to make sure wood-shingled roofs are allowed.

Shingles should be of red cedar: heartwood and edge-grained so that they won't curl. White cedar is not recommended for roofs because it tends to curl excessively. The shingles are 18 to 24 inches long with ¼-inch-thick butts. Eighteen-inch shingles must be exposed 5 inches to the weather on a 5-in-12 slope and 3¾ inches on a 4-in-12 slope. Use of wood shingles on a slope of less than 4-in-12 is not recommended. For a slope of less than 4-in-12, generally, use roll roofing. If you can find 24-inch shingles, they can be exposed 7 to 12 inches on a 5-in-12 slope and 5¾ inches on a 4-in-12 slope.

There are also red cedar "shakes." These are shingles hand-split on the exposed sides and sawn on the bottom side, with butts of ½ to ¾ inch. They come in lengths of 18, 24, and 32 inches and can be exposed 7 inches, 10 inches, and 13 inches respectively.

Preparation for wood shingles is similar to that for asphalt ones. Double the first course of shingles, extending them ½ inch beyond the eaves and rake to act as drip edges. You don't have to use a metal drip edge with wood shingles or shakes, but if you do, extend the shingles ½ inch beyond the edge of the drip edge.

Use two shingle nails in each shingle: 3d or 4d

galvanized ones are good. The nails are spaced ¾ inch from each edge and 1½ inches above the butt line of the next higher course.

Space all shingles ⅛ to ¼ inch to allow for expansion when they are wet. The joints between shingles must be 1½ inches from those in the course below. Never line up joints in one course with the joints in the course above it.

For shakes, longer galvanized nails are necessary, as is roofing felt. Use 30-pound felt, a strip under each course of shingles (and don't forget the 45-pound felt at the eave edges, to prevent leaks from ice dams). The strips can be 36 or 18 inches wide. Place a strip under the undercourse (which can be regular red cedar shingles) and the first (starter) course at the edge of the eaves. Then place a strip about halfway up the first course of shakes so that its bottom edge is well above the butt of the next course.

At ridges and hips, flashing of roll roofing or metal is applied. Make it 8 inches wide for a 4-inch drop on each side of the ridge. Shingles are applied like the Boston ridge for asphalt shingles, except that they are not folded over the ridge but rather are laid in overlapping pairs. Each pair is overlapped alternately, to prevent the ridge from having a continuous exposed joint. The open ends of the shingles should face away from the prevailing wind.

A metal ridge can be used, but even better, and in keeping with the overall wood look, is a wooden ridge made of two 1 x 4s or 1 x 5s nailed together in an L shape and then nailed to the ridge. A ridge vent would, of course, take the place of any other kind of ridge.

Flat Roofs

On nearly flat roofs and on pitches less than 4-in-12, roll roofing is recommended, although

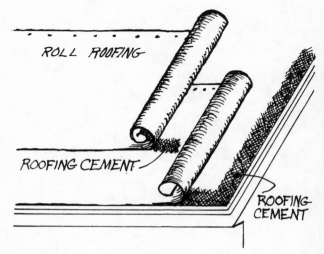

FIGURE 164. *Roll roofing is usually used on roofs with less than a 4-in-12 slope, and always on slopes of 2-in-12 or less. It is put on with roofing cement and nailed where another row will cover it. It is half-lapped.*

asphalt shingles will work quite well on slopes as gentle as 2-in-12, with the exposure kept at 4 inches. Preparation for roll roofing is the same as for the other roofs. An under layer of 15-to 30-pound felt is necessary. Then strips of roll roofing, which is heavy felt embedded with gravel and weighing 90 pounds per square, is applied and overlapped by half its 36-inch width (Figure 164). The bottom half of each strip and the edges at the eaves and rakes are secured with roofing cement. The top half is nailed; the nails are covered by the overlapping of the second strip of roofing.

If you have a completely flat roof, a built-up one is the only way to go and should be done professionally, because there is no easy way the layman can secure or rent pots for heating tar. This roof consists of built-up layers of roofing felt and hot tar. The first layer is 30- to 45-pound felt, laid dry

Roofing

and nailed on the sheathing, overlapping by 6 inches so that the hot tar will not leak through sheathing and joints. The hot tar is applied over this layer, then another layer of felt, hot tar, and so on, until four or five successive layers of felt are applied, each one separated by hot tar. The final hot-tar layer is sprinkled liberally with small stones, called gravel, to reflect heat and resist wear and tear.

Gutters

If you can avoid gutters, do so. Gutters are strictly to divert roof water away from the foundation. You probably won't need them if you have a wide overhanging roof and if your land slopes down from the addition or the house. It can help to build a concrete apron around the foundation, about 6 inches deep and 18 inches wide or wider than the overhang, to direct the water away from the foundation.

If you do need gutters or insist on them, use aluminum ones. Wooden ones look nice and are traditional, but they are expensive, need continual maintenance, and often are not big enough to accommodate all the roof water.

A 4 x 5-inch aluminum gutter is large enough for virtually any domestic building. Hangers should be put up every 24 inches in order to bear a snow load and to permit any ladder to rest against the gutter.

The longest run for a gutter is 30 to 40 feet. If the run is more than 30 feet, a downspout at each end is recommended, with the gutter sloping about $1/16$ inch per foot. Incidentally, a level gutter will drain water.

When installing gutters, extend the roofline in your imagination over the gutter so that it will

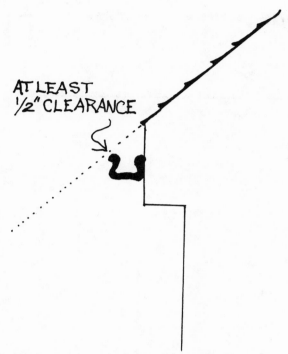

FIGURE 165. *A gutter can have a minimum slope, but its position must be accurate for good acceptance of roof runoff. The front edge of a gutter should clear an imaginary extension of the roofline by at least $1/2$ inch.*

clear the outer edge of the gutter by at least $1/2$ inch. This prevents the water from melting snow from backing up under the shingles (Figure 165). Downspouts, of the same material as the gutters, are attached to the house with strap hangers or brackets to hold them away from the walls. You can also get elbows of 45, 60, 75, and 90 degrees to allow a spout to turn a corner.

A downspout can be diverted into a storm sewer (never into a sanitary sewer), or into a dry well 10 feet away from the house. A dry well is simply a hole in the ground, 3 feet in diameter, 4 or 5 feet deep, and filled with coarse gravel (4- or 5-foot stones) or rocks from your property.

The top of a downspout should be covered with a mesh cup to prevent debris from getting into it and particularly into the dry well. The gutters also should have mesh leaf guards laid on top to prevent leaves and other stuff from clogging them. Gutters and gutter guards should be cleaned twice a year.

chapter 13

The Fun Part

Siding

Now comes the fun part—putting up clapboards or shingles. There are other kinds of exterior siding, such as stucco, brick veneer, and vertical boards or grooved plywood. There's also fake siding, such as aluminum or vinyl. These are not recommended.

You can mix or match siding: that is, match what you have on the existing house, or put a siding on the addition that is different from what's on the house. For instance, if you have a brick house, wood siding on the addition is perfectly acceptable. Even if your house is clapboard, you can put shingles on the addition, or vice versa. If the addition has a wall that is a continuation of the house, then it might be awkward to change siding, unless you put up a vertical board to separate the two units (Figure 166).

Wood siding is the easiest to put up and among the least expensive. Cedar is the most common wood. It will resist decay, last indefinitely if stained or painted, and last for twenty to forty years if not treated with anything. Untreated cedar, particularly as shingles, tends to erode when exposed to weather. But for that period of time, untreated cedar is virtually maintenance-free.

Clapboards

Clapboards, commonly known as bevel siding, are boards of red cedar tapered from ½ inch at the butt to ¼ or ⅛ inch at the top. You can get pine clapboards, but avoid them; they will split and not stand up very well, even when stained or painted. Redwood clapboards are also good, and will take paint or stain well. Avoid hardboard and other composition clapboards.

Clapboards come nominally 6 and 8 inches wide (5½ and 7½ inches respectively), and are normally overlapped one inch (Figure 167). The wider the exposure, the more contemporary the look. The one-inch overlap is recommended for weathertightness, and to permit nailing about 1¼ inches up from the butt, so that the nail will not go through the clapboard below. This is to allow the boards to expand and contract without splitting. You may hear that this is wrong, but stick to your guns and make sure the nailing is done through one clapboard thickness, not two. If you plan a small exposure (say 3 or 4 inches) and can get only 5½-inch clapboards, then you must nail through two.

Clapboards are applied over sheathing paper or

Siding

FIGURE 166. Where siding on the addition (shingles) is different from that on the house (clapboard), and the walls are on the same plane, a divider board will separate the two.

roofing felt, both breathable but not vapor-resistant, to provide a tight seal under the siding. A new material on the market, called Tyvek, made by Dow, stops the passage of air but not of water vapor; it is an excellent choice for the under layer.

You're almost ready to start siding, but not quite. You have to determine how to space the clapboards along windows and doors and under windows, to make sure they come out even. Here is how to do it. Suppose the height of the window, from the sill to the top of the casing, is 57 inches, a nice odd number. If you use a 4-inch exposure, divide 4 into 57 and get 14¼ courses. Divide 14 into 57 and you get 4.07 inches, less than 4¹⁄₁₀ inches. The fraction is so small that if you make the exposure 4⅛ inches for eight courses, and 4 inches for the other six, only you will know the difference. Figuring the spacing from the bottom of the window's sill down to the bottom of the first course is done in the same way.

Sometimes, with windows of different heights and varying distances above and below them, there is no way that you can come out even all the time. In that case, try to come out even in most of the places, and let the clapboards come out uneven wherever necessary. They will have to be notched if they do (Figure 168).

The spacing may be critical when you reach the top of the wall. Then adjustments may have to be made to the top several courses, or to the frieze board under the eaves. The point is, you don't want to end up with a one-inch clapboard under

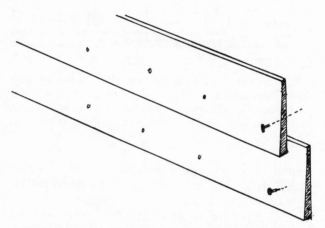

FIGURE 167. On properly nailed clapboards the nails clear the top of the clapboard below. Most clapboards are 5½ or 7½ inches wide; if the exposure is narrow, the nail must go through both clapboards.

the frieze board because it would be out of proportion.

You can install clapboards from the bottom up or the top down. To do it from the top down, nail the top clapboard at the top, and either cover its top with the frieze board or butt it to the bottom of the board. Make sure it is level. Then slip the next clapboard below into place, under the first one, and nail the top one near its butt. Slip the third clapboard under the second and nail the second. This way you can determine the exposure and check the levelness at the same time. Use a

113

The Fun Part

FIGURE 168. *If clapboards' butts do not line up with the bottom or the top of a window frame (or the top of a door frame), they must be notched to accommodate the frame (hatched area). This is to be avoided if possible.*

template to determine the exposure; if it is the same along the clapboard's entire length, it will be level. But check with a level anyway.

It is probably easier to install clapboards from the bottom up. The first course should have a layer of shingles underneath to create a drip edge and to prevent that "chinless" look along the bottom of the wall. Or you can nail a plaster lath (wood strip) along the bottom, then nail one thickness of clapboard on that.

Use 4d galvanized siding nails and drive them flush with the clapboard. The clapboard is too thin for proper setting or countersinking of the nails. You can also use aluminum or stainless-steel nails. Some stains will not cover the nails; that is a fact you must live with. If the clapboards are left to weather, you can use cut-steel nails; they will rust and make streaks along the weathering clapboards, which is a purposeful design. Make sure the nails are lined up; randomly placed nails will look terrible if allowed to rust and streak. If clapboards are extra thick (some may have ¾-inch butts), use 5d or 6d nails.

Keep vertical butt joints to a minimum. Use long boards over and under windows and over doors. Stagger butt joints so they won't line up on adjoining courses. Caulk all vertical butt joints. Caulk all joints between window and door casings and clapboards.

In Chapter 10 we suggested putting a strip of roofing felt along the side casing of each window and door. As you install the siding, the reason for this becomes obvious. Near the bottom of the window, the clapboard is installed with the strip over the board. The next clapboard goes on top of the strip, and the exposed part of the strip is cut off at the butt of that clapboard. The felt strip will divert any water that gets under the clapboards next to the window and door casings and also acts as an air stop.

SHINGLES

Shingles are another form of horizontal siding and come in red cedar and white cedar. Red cedar is generally used if the shingles are to be painted or stained. White cedar is generally used when the shingle is not treated in any way, and will weather naturally to a silvery gray color. Both will last for many years and resist decay.

Red cedar shingles are 18 or 24 inches long, with ¼- to ½-inch thick butts, and are exposed up to 8 inches, depending on their length. They are resawn and rebutted, which means they have 90-degree corners. They are kiln- or air-dried, and take paint or stain very well. Specify "perfection" or "clear grade"; anything less has knots, which you want to avoid.

There are other kinds of red cedar shingles. One is striated (grooved), is exposed 12 inches, and is face-nailed. The others are so-called

FIGURE 169. *Shingles or shakes are made of red or white cedar. Red cedar shingles are cut into many fancy shapes. From left, below: straight, pointed, diamond, round, fish-scale, and pentagon. In the circles, from left: fish-scale, staggered butt, and diamond patterns.*

shakes, which have been split and resawn, with the smooth (sawn) side designed to lie against the sheathing. These shakes are 18 to 36 inches or so long, and are laid up with large exposures. They have butts ¾-inch thick, and because of their massiveness are not recommended except for some pretty good-sized buildings or exposures. Sometimes they are set in uneven courses for an even more rustic effect. Obviously you wouldn't put them on the front of a Cape Cod–style house.

Shingles with large exposures, like clapboards, have a modern look. The smaller the exposure, the more traditional the look. You can also buy red cedar shingles in almost any shape to match those on Victorian houses: pointed or round-bottomed, or whatever, to lay up in a fish-scale pattern or with staggered butts or other ways (Figure 169). You can also double the courses so that each course has two layers: either overlapped, where the top layer overlaps the bottom layer by an inch, giving a strong shade line; or underlapped, where the top layer leaves an inch of the bottom layer exposed. In both cases, the bottom layer can be a lower grade of shingle. Of course, a double-course job would take twice as many shingles as a single-course one.

White cedar shingles come in 16-inch lengths and can be laid up with a maximum exposure of 6 inches. They are not resawn and rebutted, so sometimes they must be shaved along their sides to line up properly. Also, they are not dried, but are green, with a high moisture content, so they should be tightly butted against one another. They will shrink as they dry out "in service"; that is, installed. Normally they are not painted or stained; if they are to be, they should be allowed to dry out for six months after installation. Left to weather, they will turn an attractive gray, particularly if they are subjected to salt air at or near the sea.

When buying white cedar shingles, specify "extras," which are clear of knots and defects. Many grades of "clear" are not clear at all, and can mar a wall.

To install shingles, red or white, determine the amount of exposure and figure the height of the walls, the height of windows and doors, and the distance from the bottom of the first course

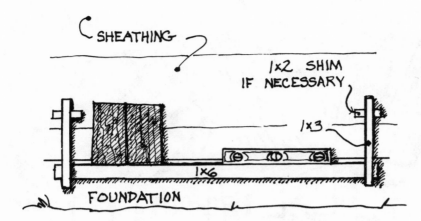

FIGURE 170. *A guide board is held at the foundation by two 1 x 2 or 1 x 3 boards nailed to each end and to the sheathing.*

to the bottom of the window sills. You will have to decide if you want to vary the exposure in order for courses to come out even under windows and along windows and doors, or to cut the shingles to fit.

Double the first course, using second-, third-, or lower-grade shingles for the undercourse. Make sure the first course is level. To do this, nail a short 1 x 2 or 1 x 3 at each end of a 1 x 4 or 1 x 6 straight board (Figure 170). Align the top of the straight board along the point where the bottom of the first course will line up and nail the 1 x 3s to the sheathing. Check the level of the straight board with a 4-foot mason's level. The bottom of the first-course shingles should overlap the foundation by ½ to 1 inch.

The doubling of the first course creates a drip edge, allowing water to drip a little way from the foundation. It also looks better; without the doubled course the bottom of the wall would have a strange, pushed-in look. For the same reason, shingles over doors and windows should be put over some sort of shimming material, usually the top half or two thirds of a shingle. This will bring the shingles out far enough to match the surface of neighboring ones that are not over doors and windows. The same point, incidentally, applies to clapboards.

Use 4d to 6d galvanized shingle nails. Shingles can be nailed directly through the sheathing and need not be nailed into studs. Small and medium shingles (up to 6 or 8 inches) are blind-nailed, that is, about an inch above the bottom of the next course above. This way the course above will cover the nailheads. Large exposures are face-nailed 1 to 2 inches from the bottom of each shingle.

Once the first course (both layers) is nailed, use a 1 x 6 or similar straight board against which to butt the shingle bottoms of the next course (Figure 171). Nail the board to the wall and make sure it is level: measure along the shingle below and mark the correct exposure at one end of the board, then at the other end. Check the board, as you did before, with a 4-foot mason's level. Holes made by the nailing of the lining board will not show.

Do not skimp while applying shingles. That is, don't try to snap or draw a level line against which to line up the next course; you will not be accurate. The board will keep the bottoms straight and will not allow you to put up a crooked shingle, which is possible because of those not-quite-90-degree corners on white cedar shingles.

Joints between shingles should be at least 1½ inches away from joints above and below. Avoid aligning joints two courses apart. The more staggered the joints, the more weatherproof the siding.

When clapboards or shingles are applied along the angle of a gable, they must be cut at the same angle as the roof. It is best to install the rake board or frieze board under the overhang before applying siding. Then cut each course of shingles or clapboards to butt tightly against the frieze board, and caulk along the bottom edge of the board before applying the siding. Another technique, not recommended, is to apply the siding right up to the overhang and nail a frieze board on top of it. The result is watertight, but it creates gaps between courses of shingles or clapboards, particularly the latter.

If you have made a half-gable dormer addition, obviously the roof angle for the dormer is different from the old one. In this case you will have to put on new siding on that part of the gable; you may want to paint or stain the entire gable wall (see Figure 8).

Vertical Siding

Vertical siding (Figure 172) can be matched boards; tongued and grooved boards 1 x 6 or greater (they can be applied horizontally, too); board and batten; spaced boards, 1 x 6 or wider, with spaces or joints covered with 1 x 2 or 1 x 3 battens; reverse batten: just the opposite of board and batten; or board on board: 1 x 6 boards with wide spaces covered by similar boards.

You can also buy plywood that is grooved to look like vertical boards; it should be stained or painted to prevent it from delaminating. Sometimes vertical siding or plywood can be installed without

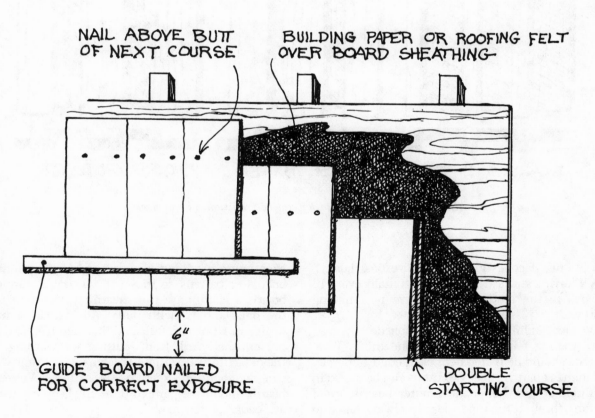

FIGURE 171. *A guide board is positioned along the second course of shingles to mark the proper exposure and to make sure the third course is level.*

The Fun Part

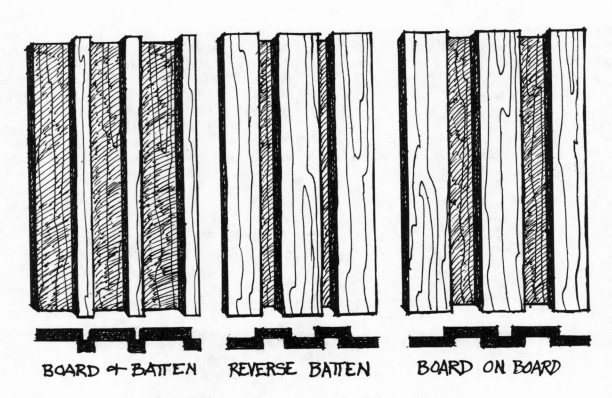

FIGURE 172. Vertical siding can be traditional or modern.

sheathing, if codes allow it. For vertical boards, 1 x 4 furring strips are nailed horizontally every 16 inches to the studs so that there is a nailing surface.

Vertical siding is normally applied, however, over roofing felt stapled to the sheathing. Wide boards should be nailed near each edge, generally 16 inches apart vertically. Battens can be nailed in a single row. When nailing outer boards, avoid nailing through the inner boards; always nail into the space, to allow for expansion and contraction. Use galvanized nails long enough to penetrate the sheathing; use longer nails (8d) for the outer boards or battens so they will go past the inner boards and through the sheathing. Countersink the nails and fill with putty. All vertical siding boards must be carefully and thoroughly nailed to prevent the boards from splitting and warping. If they start to warp, there is virtually nothing you can do about it except replace them. Vertical boards should be painted or stained, unless they are cedar or redwood.

With vertical boards, window and door frames may have to be built up to extend beyond the siding (Figure 173). This can be done by doubling the thickness of the frame or nailing band molding along the outer edges of the top and side casings, mitering the corners. Another way is to install the vertical siding before installing the windows.

Corners

Applying siding is a simple matter, but corners can pose a problem. Fortunately the problem is readily solved; in fact, there are several solutions (Figure 174).

Cornerboards are the most common, and are good with both traditional and contemporary designs. Cornerboard dimensions depend on the size and height of the addition. The higher the wall, the wider the boards. If you choose a wide cornerboard, butt a 1 x 6 or 1 x 8 against a 1 x 5 or 1 x 7 (the odd sizes may have to be cut to size) in the shape of an L and nail. The edge of the smaller board is set against the flat side of the larger one; this way the two sides of the L will come out even, or nearly so. Boards of the same size can also be used; it will probably make little or no visual difference. (Of course, the boards can be put on the corner separately instead of being made into the L shape first.) Nail cornerboards before the siding goes on, over a strip of roofing felt folded and nailed over the corner and extending beyond the boards. Siding will cover the felt. Make sure the boards are plumb, and butt all clapboards and shingles against them.

Another cornerboard technique is to nail two boards of equal width at the corner so that their inside edges (those touching the wall) touch at the corner, over a strip of roofing felt. Then fill the miniature inside corner thus created with a ½-inch or ¾-inch quarter-round, making an eleg-

FIGURE 173. *When vertical siding is installed, the window casing must be brought out by doubling the thickness or adding band molding on the top and sides of the casing. Another way to do it is to apply the siding first and install the casing last.*

ant, rounded, even-sided cornerboard. You can also buy, in some lumber supply centers, bull-nosed corner molding that looks like the above but is narrower. You can add cornerboards to fill it out.

You can make clapboards or shingles turn a corner by mitering them and caulking them where they meet. This gives a contemporary look to the siding. Or you can overlap clapboards or shingles; the overlapping is done alternately on each course. You can buy, if you look hard enough, cedar shingles cut in the form of a corner. They are expensive. And you can buy

The Fun Part

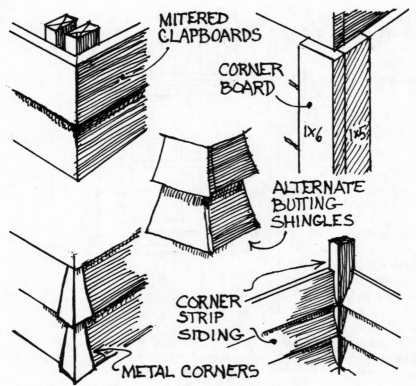

FIGURE 174. *Corner treatments for clapboards and shingles.*

metal corners, which are good only if you plan to paint or stain.

Inside corners are simpler than outside ones. Over a strip of roofing felt, nail a 1 x 1 into the corner and butt the siding against it. Or build an inside cornerboard of 1 x 3s or larger board.

If the exposure of the siding along one wall doesn't match the exposure on another wall (that of the addition, for instance), then the butts of the courses will not line up. This isn't a disaster. Separate the different sidings by installing an inside cornerboard made of 1 x 4s or larger boards. If the wall is straight, with no corner, nail a 1 x 6 or 1 x 8 vertical board separating the two sidings, and paint or stain it the same color as other trim.

chapter 14

Fussy, Fussy

Electricity, Plumbing, and Heating

Now is the time, after you have made the addition "to the weather" but before you insulate further and do interior work, to install wiring, plumbing pipes, heating and cooling pipes, and any ducts. It is also the time to string wires for burglar alarms and fire alarms, put in antennas and connecting wires for TV and FM radios, and install a central vacuum system, if you plan one.

It is not within the scope of this book to describe step-by-step methods of installing electricity and plumbing. Rather, we will show how to plan for them and consider techniques to accommodate the superinsulation and double walls.

Standard installation of wire and pipes allowed them to come into the house willy-nilly, above the ground and through the wall. Wherever they penetrated a wall, there was a potential—no, an actual—air leak. And anytime you get an air leak, you get a heat leak. With superinsulation, part of the building's ability to hold heat is that it is completely sealed, which means no air or heat leaks.

All wiring and pipes should come into the house through the basement. Through the foundation is not bad; through the floor of the basement, crawl space, or slab on grade is better (Figure 175).

Because the superinsulation of our building involves double walls and extra-thick insulation, wires can go between the walls and between insulation layers (Figure 176). Pipes could also go here (Figure 176), but it would be better to set them on the inside of the wall (Figure 177) so that there is full insulation between them and the outside. Water-supply pipes should have as much insulation between them and the outside as possible so

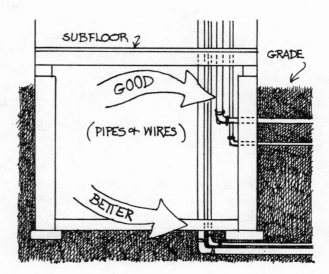

FIGURE 175. Pipes and wires should enter a super-insulated structure below grade, preferably through the basement slab. This is just another way to reduce penetration of the insulation integrity of the structure.

Fussy, Fussy

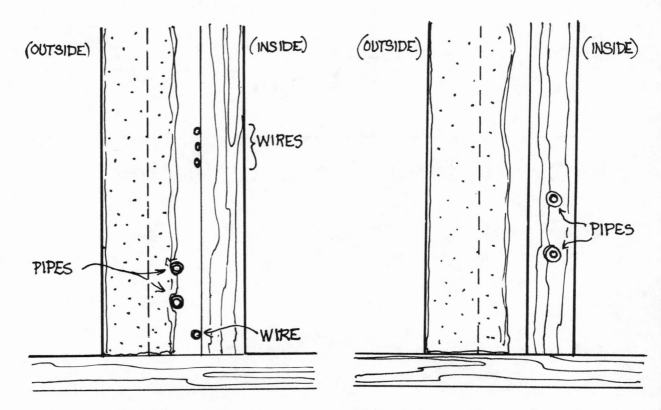

FIGURE 176. Wires can go nearly anyplace in a superinsulated double wall. Pipes can be set in the middle of the wall, between each layer of 6-inch insulation.

FIGURE 177. For maximum protection against freezing, water pipes should go as close as possible to the inside of the double wall.

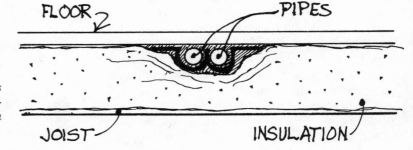

FIGURE 178. Under a floor, pipes should go close to the inside of the insulation, with a maximum of insulation on the outside.

that water will not freeze in them in severe weather. So, too, with pipes that go under the ceiling of a cold basement or crawl space (Figure 178).

WIRING

While you may not want to or cannot by law install electrical wiring and fixtures, you can save money by assisting the electrician. You may be able to hire yourself out to do all the dirty work, such as stringing and stapling cable and cutting notches in studs and joists to accommodate the cable. Laying of cable is easy in a building before insulation and wall surfaces are installed, and the very devil to do afterward. Cable needs only to be stapled where it goes along a stud or joist, or where it crosses a stud or joist.

String wires to closets, hallways, attics, basements, and stairways. Even if you don't plan lighting in these places right away, you can place the wires and install the outlet boxes with blank covers on them. They'll be there when and if you need them.

Other things to install when the walls are open are wiring for a door-bell system, intercoms, stereo speakers, fire and burglar alarms, and smoke detectors; TV and radio antennas; and, with the help of the telephone company, telephone wire, with outlets in each room for either permanent phone installation or a jack for a portable phone. This is also the time to put in the ductwork for a central vacuuming system. Finally, it might be good to place the wiring for room air conditioners. You may not want or be able to put in all these things at this time, but in the future all you have to do is buy them and literally plug them in.

The location of electrical outlets and switch boxes is important to the integrity of the vapor barrier. Outlet boxes should be spaced every six feet along all walls. Plan for their placement now, even though, on outside walls, the boxes should be mounted *on* the wall, not in it (Figure 179); much of the work in installing them is done when putting up wall surfaces (Chapter 15). It is particularly important to put the boxes on the wall when there are only 6 inches or so of insulation on the wall. Setting the boxes in such a wall will reduce the insulation behind them and will break the vapor barrier, creating, again, an air leak. For a doubled wall with two layers of 6-inch insulation, this is less critical, but it's still a good idea to put the boxes on the inside of the interior wall. It doesn't matter, of course, on inside partition walls, except for sound control, in which case boxes should not be lined up on each side of the wall but rather offset by at least 18 inches.

Outlets are put 12 to 14 inches above the floor. In kitchens, bathrooms, and laundries they

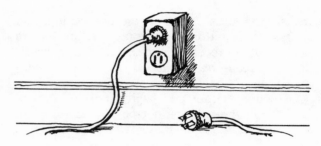

FIGURE 179. *A face-mounted outlet box. Setting electrical boxes on the face of a wall, with only the cable penetrating the vapor barrier and wall, instead of fitting entire box into the wall, provides maximum insulation and minimum disruption of the vapor-barrier skin.*

should be located for easy access, which means 36 inches above the floor or a few inches above any countertops. A dining room should have at least one outlet high on the wall (36 inches up or so) so that appliances can be used at the table or on a sideboard. Kitchen outlets should be spaced more closely than those in other rooms.

Switches should be located where they can operate ceiling or wall fixtures. Multiple switches should be used where there is more than one entrance to a room, and on stairways, allowing the light to be turned on or off at either entrance or at the bottom or top of the stairs. They are set 48 inches from the floor.

In most rooms, outlets can be of the split-control type: one outlet is permanently live, the other controlled by a wall switch so that a room with no ceiling fixture can be lit by a table or wall lamp plugged into the outlet controlled by the switch.

Avoid recessed lights in ceilings that are below an attic. Such a fixture must be free of insulation on the attic side (for at least 3 inches) to prevent buildup of heat. If you want a ceiling fixture,

Fussy, Fussy

make sure it is mounted on the ceiling, not recessed. Junction boxes and ceiling-fixture boxes that are recessed into the ceiling, however, do not create heat and can be covered with insulation.

Of course, outlet and switch boxes mounted directly on the wall cannot be affixed until the wall surface is installed; see Chapter 15.

Plumbing

Plumbing is another matter, because it has to be installed in walls and basement ceilings and must be protected so that water in the pipes won't freeze. So install water-supply pipes (plastic or copper — copper pipes are preferable because they pose no fire hazard) toward the warm part of the building, with all the insulation between them and the outside. As we've seen, the problem of freezing is avoided if the pipes are put in the interior walls of the double-walled superinsulated wall system.

In a basement or crawl space, pipes traditionally have gone under the joists, but with heavy insulation in the basement ceiling, keeping basements cold, there is a risk of freezing. So the pipes should be near the warm side, right under the floor, with the insulation between them and the basement. This is especially important for pipes in a crawl space (see Figure 178).

Plumbing should be carefully planned. Try to group rooms (laundry, bath, and kitchen) so that pipes, drains, and vent pipes can be located in one area. Drainpipes are 3 or 4 inches in diameter, and the 3-inch pipe can barely be accommodated in a 2 x 4 interior wall. A 4-inch pipe requires a 2 x 6 wall (Figure 180) and it wouldn't hurt to install a 2 x 6 wall with 3-inch pipe as well.

FIGURE 180. *A 4-inch plumbing drainpipe requires a 2 x 6 floor plate, studs, and top plate.*

Planning will also enable you to keep from cutting too many floor and ceiling joists. Joists should be cut only where the effect of a decrease in strength due to cutting is minor. Where it is required, reinforcing plates of plywood running 2 feet on either side of the cut can assure full joist strength. Holes should be made only in the end quarters of a joist, not less than 2½ inches from the top or bottom of the joist, and not more than 2 inches in diameter. If larger holes are required, or if they must be nearer the top or bottom of the joist than the 2½ inches, reinforcing can be done with either plywood plates or headers between joists.

The only other kind of reinforcing is needed for bathtubs, particularly cast-iron ones. The cast-iron tubs are heavy; so are others when full of water *and* a bather. To prevent sagging of floors, therefore, joists at the outer edge of the tub, or at the end if the tub is at right angles to the joists, should be doubled. And when needed, the intermediate joist under the tub should be offset to allow for the drainpipe. The wall along which the tub sits should have extra studs nailed to the regular studs, cut short and offset slightly (brought into the room a bit) so they can support the lipped edge of the tub (Figure 181). Regular studs should have blocking or sleepers between them just above the edge of the tub to act as nailers for wall surfaces.

Heating

Whatever kind of heating you plan—hot air or hot water, fired by any fuel—getting the heat to the rooms is critical. It is important to make sure hot-water pipes or air ducts are located properly so they won't interfere with the integrity of the

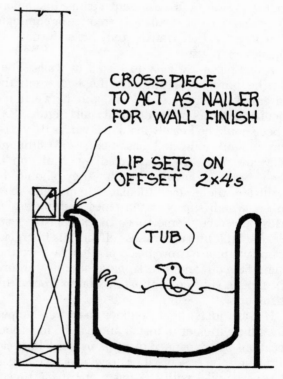

FIGURE 181. *Short studs are set out slightly beyond regular wall studs to hold the edge of a tub. Filler pieces between studs are nailers for wall finish.*

addition's wall and basement ceiling insulation. They also must be insulated to reduce or eliminate heat loss in their transit from boiler-furnace to radiator or register.

Water pipes are easy to install in walls. They can be brought up through the floor from the basement or crawl space and put between the insulation layers in the doubled wall (Figure 182). They should be heavily insulated where they pass through an unheated basement, and although they are pretty well protected by insulation in their passage between walls, they should be insulated there, too. Insulate them with foam tubes, which slip over the pipes and are sealed with duct tape. Some tubes have a plastic seam closure built into them. Less effective is fiberglass spiral wrap, a narrow piece of fiberglass wound, spiral-fashion and overlapping slightly, around the pipe. A vinyl cover must be wound around this insulation.

Hot-air ducts, being bulkier than water pipes, are a little trickier to install and can be in various locations. An expanded plenum, or trunk line, can run along the basement or crawl-space ceiling. If it runs parallel with the joists, it can be tucked into the space between them (Figure 183). In fact, in some cases the space itself can be used as a trunk line and sometimes as a cold-air return (Figure 184). The two are, of course, separate. Insulation should be applied to the bottom of such parallel running ducts. Where the trunk line runs at right angles to the joists, it naturally would go under them (Figure 185).

Some branch ducts are 5- or 6-inch round pipe; others, including risers that go up the wall to the second floor, are 3 by 10 inches, and will fit in the inner portion of the double wall (Figure 186).

In all cases, ducts should be insulated with 2-inch fiberglass duct insulation, which is wrapped around the duct or pipe and stapled in place

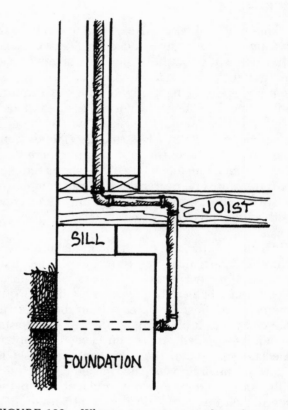

FIGURE 182. *Where water pipes go through an unheated basement or crawl space, they must be heavily insulated. This is particularly important if the basement or crawl-space ceiling is insulated, which will make the basement even colder.*

Electricity, Plumbing, and Heating

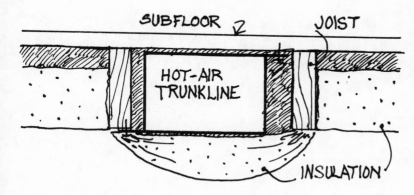

FIGURE 183. A hot-air trunk line can be tucked in the space between joists, and insulation secured below it. This eliminates the need to insulate the trunk line itself.

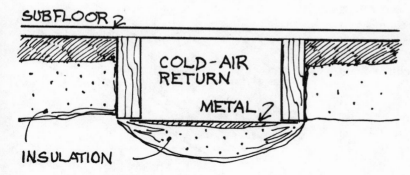

FIGURE 184. Cold-air returns should be treated the same way as a hot-air trunk line.

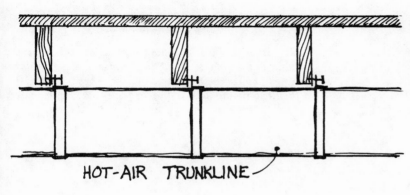

FIGURE 185. If the trunk line is at right angles to the joists, it must be hung below them. In such cases, the trunk line itself and the space between the joists should be insulated.

Fussy, Fussy

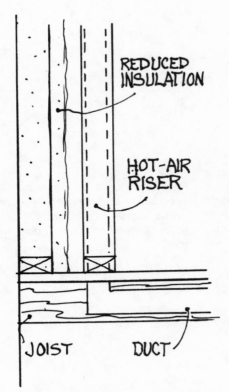

FIGURE 186. *A duct riser fits between studs; in this case it should be the inside unit of a double wall.*

(Figure 187). The duct insulation is faced with a vinyl skin, which must be on the outside. Even better, but awkward to install, is 3½-inch fiberglass insulation with paper or foil skin. Another type of duct insulation is a cellular foam material with a foil skin and a self-adhesive back (covered with peel-off paper).

Since your addition is superinsulated, chances are that you won't need an elaborate heating system, particularly if you use a wood or coal stove or other type of heat to keep cozy.

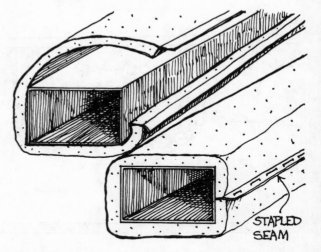

FIGURE 187. *Two-inch duct insulation is wrapped around hot-air ducts, and the seam stapled. Regular 3½-inch house insulation can also be used, but is difficult to install, and its foil or paper backing is more delicate than the vinyl skin of duct insulation.*

chapter 15

The Smoothies

Walls and Ceilings

Insulation and vapor barriers are installed; wiring, pipes, and ducts are in place. It's time to put up wall surfaces: plaster, plasterboard or wood boards, or paneling.

In the good old days, one could afford to put up plaster walls. Plaster was applied on wood or steel lath, first in a rough, or brown, coat, then in a thinner, smooth, or finish, coat. It is expensive now, but is among the best wall finishes possible. It is not for the do-it-yourselfer. Today's plastering technique is different from yesterday's only in the lath. Instead of spaced wooden strips, which allowed the plaster to ooze into the gaps to be keyed into place, Rocklath is used — 3/8-inch-thick pieces of plasterboard nailed to the studs.

Such walls are rare today because of the low cost of installing sheets of plasterboard to the walls, also known as dry wall (Figure 188). There are two types: (1) plasterboard nailed or screwed onto the wall with the joints and screw or nailheads covered with a plasterlike material called joint compound; or (2) a special moisture-resistant plasterboard called Blueboard, applied to the walls and the entire surface covered with a skim-coat of chemically setting plaster. The latter is not a do-it-yourself job, either, but is a better wall than the first one.

The two types of plasterboard are applied in the same way. Start with the ceilings. In order to make a nailing surface along the edges of and across the middle of the ceiling and to facilitate the positioning of the plasterboard panels, nail 1 x 2 or 1 x 3 furring strips at right angles to the joists, over the ceiling vapor barrier. Since the joists are on 24-inch centers, put up the strips every 12 inches (Figure 189). The strips can be shimmed with shingles to ensure that they are level.

Where joists are parallel to the wall, there may be no place for the furring strips to be nailed, so an extra ceiling joist should be installed. On a partition wall that is parallel to the ceiling joists, short sleepers that match the depth of the joists are put up to connect the joists on each side of the wall (Figure 190).

For the ceiling, 3/8-inch plasterboard is the thinnest you can get and the lightest to handle, but even so, a 4-by-8-foot sheet weighs 48 pounds. The longer the sheets (you can sometimes buy them up to 10 feet long), the more weight to juggle, but also the fewer the seams you will have to fill.

To erect ceiling sheets, a husky crew can hold up each sheet while it is nailed. You can also build a T-shaped brace a little more than ceiling height that goes under the sheet to hold it while you nail. Even with a brace, putting plasterboard on a ceiling is difficult.

The Smoothies

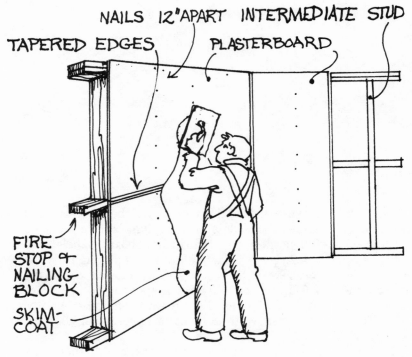

FIGURE 188. *Plasterboard can be applied horizontally or vertically. Joints are covered with joint compound and paper tape, and nailheads with compound. A new method is not to fill joints or nailheads, but to give the entire surface of Blueboard (a special plasterboard) a skimcoat of real plaster. Joints are filled with a nylon mesh instead of paper tape.*

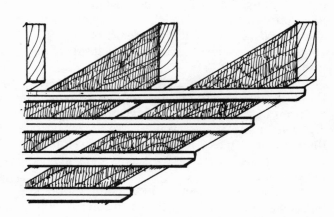

FIGURE 189. *Furring strips are placed on joists 12 inches o.c. to allow plenty of nailing surfaces for ceiling plasterboard.*

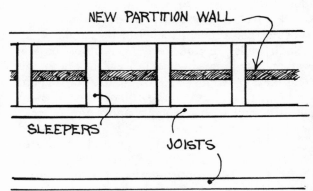

FIGURE 190. *If a partition wall is parallel to the ceiling joists, sleepers connecting the joists above the wall will act as nailers for furring strips.*

Nail the sheets with plasterboard nails: blue steel ring-shanked nails with large heads. Nail every 6 inches along the edge of each sheet, and on intermediate joists or strips nail every 12 inches. Dimple the nailheads with an extra blow of the hammer. These dimples will be filled with compound.

You can use dry-wall screws instead of nails. They hold better and will not tend to pop out if the wood holding them dries out. They have Phillips heads (crossed grooves instead of one straight slot) and must be driven with a variable-speed drill with a special bit. Because these screws hold so well, they can be spaced every 12 inches on edges and on intermediate nailing surfaces.

To determine where the intermediate strips or joists are located, mark their positions on the walls before putting up the ceiling sheets, and at each end of the sheet after it is held with a minimum number of nails or screws. Then you can connect these marks with a pencil or chalk line so that there is no guessing where to nail or drive screws. Too much guesswork (and missing of the nailing surface) will create a lot of unnecessary holes that are difficult to fill with joint compound. It would be easier to cover them with a skimcoat.

Butt the long edges of each sheet together. They are tapered so that they form a valley at the butt joint. When you apply joint compound to fill this valley, the compound can be smoothed off and tape applied without creating a hump along the joint. Where ends butt, or where you have to cut an edge and there is no taper, leave a gap of 1/8 inch at the joint. This will act as a key when filled with joint compound and also allows an adequate amount of compound to be applied with a minimum of hump.

For walls, the plasterboard can be 1/2- or 5/8-inch thick. Because studs are on 24-inch centers, the 5/8-inch is better, spanning the 23-inch gap without sagging or bending. Sheets go up the same way as on ceilings. It is best to apply the longest sheets horizontally (see Figure 188), butting the lower sheet against the floor and cutting the upper half to fit under the ceiling. Ideally, the top half is put up first and the bottom half cut and fitted against the floor, since any gap at the floor will be covered by the baseboard. But unless you have a crew to help, it's tough to get the top sheets up; you can nail a ledger to hold them temporarily in place. Horizontally placed sheets will have fewer seams to cover than sheets placed vertically. Horizontal seams should line up along the nailing blocks and fire stops you installed earlier in building the wall (see Chapter 6).

Notch the plasterboard sheets to fit above door openings and above and below window openings, to avoid vertical seams at some of the window and door corners. This is where there is the most stress when the addition settles, and a solid piece of plasterboard will have less tendency to crack than a plastered seam.

To cut plasterboard, score the finish paper with a utility knife. Then break the plasterboard at the score mark and cut the paper on the back side where it folds. Plasterboard will break and cut cleanly. Avoid sawing; it will create much dust and will dull the saw blade quickly.

As we pointed out in Chapter 14, outlets and switch boxes on outside walls should be mounted on the face of the wall to prevent, as much as possible, piercing of the wall surface.

The wire to a box need not be face-mounted; it is inside the wall. The hole through which it goes can be made as small as possible. You should caulk around the wire at this hole with silicone or other long-lasting caulking.

Boxes mounted in the walls, however, as in partition walls, must have their outer edges flush with the wall surface so that the cover plates will

The Smoothies

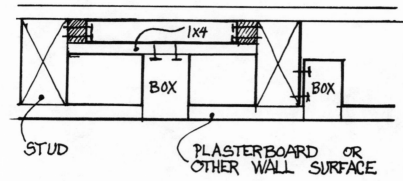

FIGURE 191. *If outlet boxes and other electrical units are set in the wall (it's okay to set them in interior walls), their faces must be flush with the wall surface. Boxes can be mounted between studs, using a 1 x 4 board or steel mounting flange, or next to the stud, with the box nailed directly to the stud.*

fit properly (Figure 191). Therefore they are placed as far out from the studs as the plasterboard or plasterboard plus paneling is thick. Boxes can be located next to a stud and nailed to it or set between studs, with a board or special steel arm spanning the stud space behind. Flush-mounted boxes should be caulked thoroughly all around the outside (do not caulk on the inside).

You can buy steel templates to place on the boxes and press the plasterboard on them, automatically lining them up for cutting. Or you can line the panel up with the box and use a square to extend the horizontal edges of the box to the plasterboard panel. For vertical edges, make two measurements: one from the point where the edge of the plasterboard will go to the near side of the box; the other, to the far side of the box. Transfer all these measurements to the panel, being sure to mark both top and bottom as well as sides so that the corners are square.

Connect all marks and cut the box hole with a utility knife or keyhole saw. Cut it ⅛ inch larger on all four sides. The box plate will cover this enlarged opening. You must also cut out notches for the ears of the box — the little studs at the top and bottom that are threaded to receive the bolts that hold the outlet or switch itself in place.

Another technique for mounting wall plasterboard is gluing. Use building or construction adhesive, which comes in cartridges that fit a caulking gun. A bead of adhesive is applied to the studs and the plasterboard pressed into place. Sometimes nails are applied sparingly to keep the panel in position while the adhesive sets. The advantage to gluing is the reduction or elimination of the numerous nailheads that must be dimpled and filled with compound. It also means there won't be any nail-popping if the wood shrinks. Glue cannot be used, however, over a polyethylene plastic vapor barrier.

Once panels are up, the bull work is done and the fine-artwork starts: filling seams and nailheads (Figure 192). Joint compound eases this job. The best to use is premixed. The compound is used alone over nailheads and with 2-inch paper tape over joints or seams. Some joint compounds can be used without the paper tape, but you should use it, regardless. The tape prevents the compound from cracking.

Apply the compound with a smoothing knife: a 5-inch knife is good for the initial application. Put the compound over the nailheads and smooth it with one swipe of the knife. Let it dry overnight and do it again, with a thin layer. Let it dry and do it a third time. If you don't do it three times, you are likely to have a slight indentation that will show up whether the wall is painted or papered.

For seams, apply the compound thickly, then smooth off with a 10-inch knife. Apply the tape, press it into the compound with the knife, and smooth. Then cover with a skimcoat of compound. The 10-inch knife will automatically feather the edges of the compound, making it superthin at the sides. Again, do this three times, feathering the edges as you go along. A secret to the success of this procedure is to wield the knife in long, steady strokes. Don't paint with it; you will only make ridges. The idea is to make the compound as smooth as possible, reducing the need for a lot of sanding or wet-sanding. Another secret is to make sure you use a 10-inch knife; anything narrower will make the job much more difficult and make it look terrible.

If you get too thick a coat near the edges, remove the compound and start over.

When all three coats are dry, sand or wet-sand or use a wet sponge to smooth the final surface.

In inside corners where wall meets wall and ceiling meets wall, the application is similar, although there is less space for the compound.

Walls and Ceilings

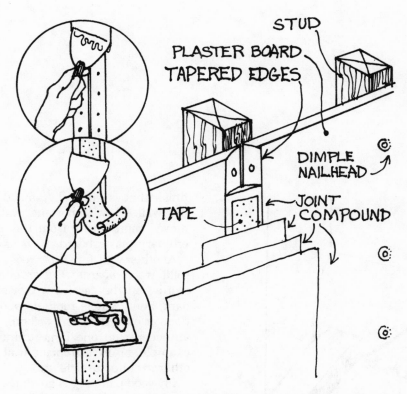

FIGURE 192. *Filling a joint in plasterboard, where the edges are tapered to form a valley. Tape is applied to joint compound in the valley, and three coats of compound are smoothed over the tape. Each coat of compound must dry before each succeeding coat goes on. Nailheads are dimpled and filled with three coats of compound. Anything less than three coats is likely to leave a depression.*

The paper tape is folded lengthwise into the shape of an *L* (Figure 193) and pressed into the compound. Then three coats are applied to the corner, one at a time after the preceding coat has dried. Each succeeding coat is smoothed and feathered, just as in the straight seam. One technique to follow is to apply the compound over the tape only on one side of the corner first and let it dry. Then do the other side. The reason is to get the corner as straight as possible. Trying to do both sides at once is almost impossible without a lot of practice; the knife keeps wavering into the other side of the corner. Corner knives are available, but they are expensive and there's no guarantee that you will be very good at using them.

On an outside corner nail a steel corner bead (an *L*-shaped strip) in place and apply the compound; as you smooth it, the corner bead will automatically taper and feather the compound. Or you can apply a wood corner bead, which can be painted the same color as the wall or painted to match wallpaper or woodwork.

The installation of Blueboard uses a plaster that sets chemically rather than by evaporation, as does standard plaster. The water-resistant Blueboard is skimcoated (about 1/8 inch). No tape is put over the joints; rather, a nylon mesh is glued or stapled into the joint valleys. This mesh is essentially reinforcing, just as steel mesh or reinforcing rods are in concrete; and it stops or prevents cracking, just as reinforcing steel stops or prevents cracking in concrete. The skimcoat goes over the entire surface: Blueboard, joints, and nail or screw heads. Its advantage is a hard plaster

The Smoothies

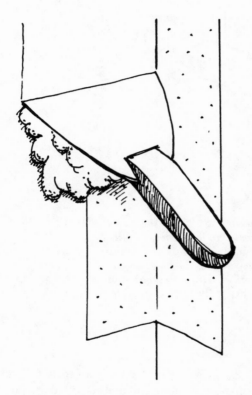

FIGURE 193. *In a corner, tape is folded into an L shape and embedded into the compound. One side of the tape is covered with three coats of compound, then the other side. This is to assure a straight corner.*

finish that is generally smoother than the older dry wall. Most important, wallpaper can be removed more easily from this surface than from ordinary plasterboard, with its paper facing.

Another wall finish is wood, either plywood or solid wood boards. Either one is effective and sturdy, but be careful with wood; it is relatively dark, and a whole room of wood can be monotonous or overwhelming or both. Consider wood as an accent; on one large uninterrupted wall, for example, or on a fireplace wall or one with some other architectural feature.

Plywood paneling is good; it's usually hardwood veneer over thin plywood, about 3/16 to 1/4 inch thick. Cheaper and really not worth the investment is the paneling made from hardboard with a photographic covering of so-called wood grain. Paneling serves a good purpose, but try to use real wood. And since it is so thin, it should not be applied directly to studs because it will tend to make a hollow sound when knocked or hit and will bend when leaned on. Best to glue it to plasterboard, using panel adhesive (similar to construction adhesive). This adhesive comes in a caulking cartridge and is applied to the plasterboard in zigzags. It will hold indefinitely. One good thing about paneling on plasterboard is that the joints and nailheads do not have to be finished off with compound. When applying plywood paneling to plasterboard, make sure the joints in the panels do not line up.

If for some reason you do have to put plywood paneling on open studs, do this: apply horizontal furring strips (1 x 2s) to the studs on 12-inch centers. That way you will get smaller gaps for the plywood to cover, and nailing can be done in the grooves of the panels.

Solid board paneling comes 3/4 inch or less thick, in soft wood and hardwood (very expensive), and usually in nominal 6-, 8-, and 10-inch widths. The

Walls and Ceilings

FIGURE 194. *One form of diagonal board paneling.*

boards can all be the same width or can be random widths. They are tongued and grooved and blind-nailed; that is, nailed through the tongue with 8d or 10d finish nails.

Horizontal solid boards are placed directly on the studs. Vertical boards are placed on horizontal furring strips, 12 to 24 inches on center. Boards can also be applied diagonally (Figure 194).

The only other wood treatment for a wall is wainscoting, which is partial paneling along the wall from the floor up 2 to 4 feet. There is no set height for it; it is a matter of design, style, and taste. Trim for wainscoting is a 1 x 2 with a band molding or similar molding above and below the 1 x 2. Actually, the molding or any wainscoting trim is limited only by your imagination.

chapter 16

Made for Walking

Flooring

You've had a floor in your addition since you put plywood or boards over the joists; something you could walk on without falling between them. Now you have to consider finish flooring. There are several kinds: wood strips, planks, parquetry, and parquet tiles; resilient (plastic) tile; carpeting; and ceramic tiles.

Flooring, and underlayment (plywood and such) for tile, are installed directly on the subfloor (rough floor) and before any interior trim, such as door casings or baseboards, is installed. This is to avoid the need for intricate cutting around casings and to allow baseboards to cover gaps at the edges of the floor.

Wood

Wood is still just about the best thing you can use for flooring. It is durable, good-looking, and can be renewed and refinished. It is installed in strips or boards (Figure 195) of hardwood, usually oak. The boards range from about 2¼ inches to 5 or 6 nominal inches wide and are tongued and grooved, as well as end-matched, which means that their ends as well as their long edges are tongued and grooved—a device that allows you to place ends of strips between joists. Oak flooring comes in various grades, from clear to somewhat knotty and blemished, which has more character.

Before installing strip flooring, put down sheathing paper or roofing felt, overlapping it by a few inches. This prevents dust from filtering through the subfloor and into the basement or other rooms and also serves as an air stop. Strip flooring must be installed at right angles to the floor joists so that each nail will go into a joist. This presents no problem when the subfloor is plywood or has been nailed diagonally. But if the subfloor boards are at right angles to the joists, a second, parallel layer will tend to cup and not behave very well despite thorough nailing, because nails don't hold very well in just the subfloor. That's why it is best to lay the subfloor diagonally.

Because wood expands and contracts with moisture in the air, the wood strips should be brought into the room in which they are going to be installed at least two weeks beforehand; this permits their moisture content to equalize with the moisture in the room. Even with this precaution, the first strip must be laid ½ inch from the wall to allow for expansion and contraction. Without this gap (and a corresponding one at the opposite wall) the strips would buckle when they expand.

Face-nail the first strip, driving the nail through the face of the board. You may have to predrill nailholes because oak tends to split when nailed.

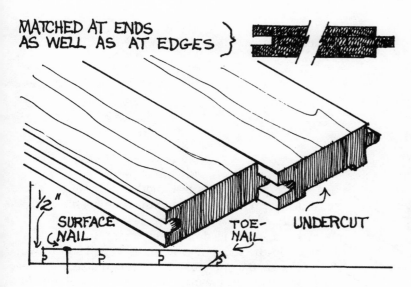

FIGURE 195. *Strip hardwood flooring is tongued and grooved along the edges; ends are also matched (tongued and grooved), so that boards don't have to be butted over a joist. When starting a strip hardwood floor, leave a ½-inch gap on all four sides to allow for expansion and contraction.*

Use flooring nails—cut nails made of steel, long enough to go into the joist for about an inch. Flooring nails have blunt ends and tend to crush their way through the wood. Pointed nails tend to separate the wood fibers, causing splits. The first strip is also nailed through the tongue, and the nail set flush with it. A special nailing tool can be used for speed and convenience. The grooved side of the next strip is then forced onto the tongue of the strip in place, and the tongue of the second strip nailed.

Sometimes a strip will be bowed. Then you must use brute force to get it up against its neighbor. You can use a spare strip as a striking surface. If that doesn't work, you can also cut a scrap board to use as a cleat and a wedge board to force the bowed board into place (Figure 196). Once it's nailed, it will stay in position. The last strip to be nailed must be ripped (cut lengthwise) to fit. It also must be half an inch from the wall. This gap is less critical at the ends of the strips because the strips expand and contract less along their length than across their width. The last strip is face-nailed. The baseboard, on top of the finish flooring, plus the quarter-round shoe mold at the bottom of the baseboard, will cover any gaps and will also cover exposed nails (see Chapter 17).

Most strip oak flooring comes unfinished, so it must be sanded, mainly to smooth away any unevenness on the boards. Other flooring, such as parquet tile and some plans, comes prefinished and probably waxed as well.

Rent a floor sander and edger. The floor sander is a monster machine with a drum, sometimes two, to which sandpaper is attached. Sand three times, with coarse, medium, and fine sandpaper. The most important thing to remember in sanding is to keep the machine moving at all times; that spinning drum will gouge the floor or make it uneven. The floor sander has a lever, or is counterbalanced, to lift the drum off the floor. When starting a run parallel to the strip flooring, move the machine and then lower the drum to the wood; lift the drum before coming to the end of a run.

Start at one end of the room (Figure 197), with your back to one wall, and walk the machine down the length of the room, following the direction of the strips, until you reach the opposite wall. Return, walking backward. The coarse sandpaper will do a quick job; repeat if necessary, until the wood is stripped to an even plane. Then move to the second pathway for sanding and do that, and so on. Then turn around and do the strips you could not reach in the previous direction. An edger, with a disk attachment, does edges, and a hand scraper is used in corners. When sanding, buy more than enough sandpaper in the three grades to do the job; nothing is quite so frustrating as to run out of sandpaper in the middle, particularly on a weekend. You can return any unused paper and get your money back.

Oak flooring also comes in tongued and grooved planks. It can be blind-nailed in the tongues if it

Made for Walking

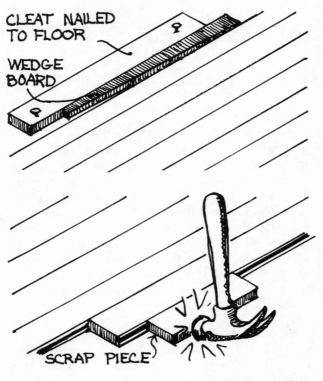

FIGURE 196. *A scrap piece of flooring (bottom) is used as a block to snug board to its neighbor. A cleat (top) nailed to the floor holds a tapered wedge board driven against its neighbor.*

isn't too wide; otherwise it must be face-nailed and the nails countersunk and filled with putty before finishing, if finishing is necessary. Usually planks are prefinished. They can also be screwed, with wooden plugs covering the countersunk screws.

With a little imagination and a fair amount of work, you can also put down oak strips in a parquet pattern (Figure 198).

Wide pine boards are another way to go. The wider the better; at least extra-wide boards give a "Colonial" look. Wide pine is not tongued and grooved, so it is doubly important to store this wood in the room in which it is to be laid for two weeks so the moisture can be equalized. If the floor is laid in the winter, when the heat (if any) is on and the boards are dry, they will be at their smallest when laid, so when they do absorb moisture and expand, they will butt even more tightly against one another. Eventually the joints will open up, but it will take years.

FIGURE 197. *Strip flooring is sanded with a floor sander in two steps: do one side of the floor, then turn around and do the short side. Three grades of sandpaper must be used whether the floor is new or very much used.*

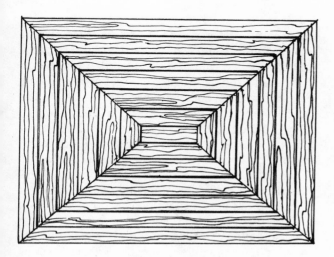

FIGURE 198. *Oak strips can be applied in a parquet pattern; for example, concentric squares.*

Install pine boards the same way you do the oak strips. Use 10d nails. They can be screw nails, ring-shanked, or hot-dipped zinc galvanized box nails. (The galvanized nails are the best.) Countersink them all. The holes can be filled with putty, but since pine boards are somewhat informal, they can be left unfilled and still will look good.

If pine boards are to be stained, make sure the edges as well as the face are stained so that when the boards shrink and joints open up, the edges will not show light against the face.

Floors over Concrete

Wood floors can be applied over concrete slabs. It is helpful if the slab is poured with a plastic vapor barrier under it. If so, 2 x 4 sleepers are nailed, wide side down, with concrete nails right into the concrete, on 12-inch centers (Figure 199). Construction adhesive can also be used. Then strip flooring is installed right on the sleepers; the 12-inch centers provide a gap narrow enough so that you will get no sagging despite the lack of a subfloor. If tiles or carpeting is to be applied, ⅝-inch plywood is put on the sleepers. Polystyrene insulation, 1½ inches thick, can be put on the floor between the sleepers.

If there is no vapor barrier under the slab, then some moisture-proofing is necessary. Hot tar, applied by a roofer, works the best. The next best is roofing cement. Then the sleepers are embedded in the tar or cement and the floor laid.

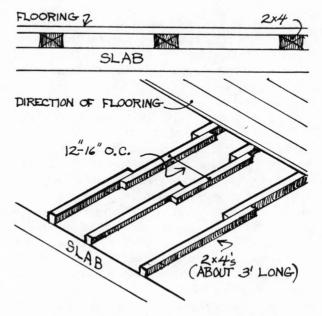

FIGURE 199. *Short 2 x 4s are laid on a concrete floor and plywood or finish flooring applied directly to them. If there is no vapor barrier under the concrete, hot tar is poured on the floor and the sleepers embedded in it.*

One thing about using concrete nails. They are hardened steel and will hold in concrete, but when nailing through wood into concrete, do not try to countersink the nails or even drive them flush with the wood; they will tend to break up the concrete. Instead, stop when the nailhead meets the wood. You can rent or buy a small tool that eases driving of concrete nails. Some people can't drive a concrete nail to save their souls. For those people, construction adhesive will work.

Carpeting and Tile

Resilient tiles and wall-to-wall carpeting make good flooring surfaces, but chances are they will not last as long as oak or even pine floors. For carpeting, the best solution is to put it down over oak flooring. The reason we suggest this is that carpeting will not last the life of the mortgage, and after it has been replaced in ten or fifteen years, you will still be paying for it. Once it's removed, you may find the oak floors or pine floors attractive enough to expose, with area rugs.

Tile needs an underlayment, an extra sheet of plywood flooring to make sure the floor is strong enough to support people and furniture. Tile (or carpeting, for that matter) laid on just the subfloor results in a springy, unsatisfactory floor. Use underlayment plywood on top of the subfloor. This is important because underlayment plywood does not have any hidden knotholes. If there were knotholes under the top veneer, small heels would break right through the tile or carpeting and veneer into the holes.

Apply underlayment plywood with ring-shanked underlayment nails, every 6 inches along the edges and every 12 inches on intermediate joists. The nails must go into the joists. If that is not possible, the plywood should be screwed to the subfloor. Joints, holes, and other defects or openings in the underlayment must be filled with a floor-leveling compound, a plasterlike material that can be smoothed off and sanded when dry.

Resilient tile is made of vinyl-asbestos (the asbestos is safe because it is bound up in the vinyl), solid vinyl, or no-wax vinyl. Tiles come in 12-by-12-inch squares and are 1/16 inch thick (commercial tile is thicker). Different kinds of tile are put down with different kinds of adhesive. When laying tile and using adhesive, follow instructions to the letter, and make sure the right amount of adhesive is applied. A notched trowel assures this, but the notches must be the right size. Self-adhesive tiles are another good way to go; they are convenient and the adhesive sticks very well.

Because the tiles are square, they are installed so that each tile is square to itself and its neighbor, not to the room. This is important if the room is out of square (corners not at 90 degrees). So tiles are not laid beginning at one wall; they are started in the center of the room (Figure 200). To do this, find the center of one wall and mark it. Find the center of the opposite wall and mark it. Then connect the two marks with a snapped chalk line or a straight pencil line. Do the same with the other two walls. If the walls are reasonably parallel to one another, the place where the lines cross will be the center of the room and the angle will be 90 degrees.

The center mark will permit you to start the tiles in the center of the room. It will also permit you to determine the width for the edge or border tiles, which must be at least half the width of a full-sized tile. Suppose the room is 12 feet wide and 12 feet long. This might sound great; you won't have to cut any tiles. Well, rooms don't actually work out that way, and because no room is dead square, you may have trouble making the

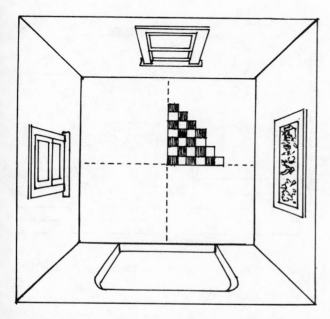

FIGURE 200. *Floor tiles are begun in the center of a room because tiles are laid square to each other, not to the room, which may not be square, or may not have 90-degree corners.*

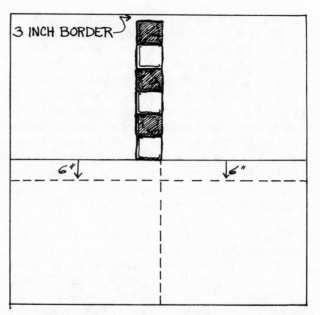

FIGURE 201. *Apply the first row of tiles dry; if the border tile is too narrow, one line is moved 6 inches; this will automatically adjust the border tile width.*

1-foot-square tiles reach the edge of the floor. So, move one of the center lines 6 inches to either side (Figure 201). This will divide the width into one section 5½ feet wide and the other 6½ feet, which allows 6-inch border tiles.

The same can be done with a width of 12 feet 4 inches. Instead of a row of 12 tiles and two borders of 2 inches each, moving the center line 6 inches allows a row of 11 tiles (11 feet) and 8 inches for each border tile (16 inches), which comes out to 12 feet 4 inches. The technique is easier done than said; once you do it you will find it's easy.

Line up the first row of tiles, dry, to make sure all this calculation does indeed work. Then apply the adhesive and line up the first tile on one of the lines (the adhesive is applied right up to the line). Don't worry if the tile does not butt up to the cross-line; if you follow one line the tiles will be installed in the best way possible. Butt succeeding tiles tightly; there is no room for fudging. Don't try to push a tile into place once it's in the adhesive; that will just plow up the adhesive and make it ooze through the joint. Lift the tile and set the edge against the edge of the tile in place, and drop it into position. Self-adhesive tiles are easier to work with, but make sure they are in the right position when you lay them. They are very difficult to lift once they are glued down.

Almost any tile can be cut by scoring it with a utility knife and breaking it at the score mark. To cut edge tile, place the tile to be cut directly on its neighbor, the last full-width tile next to the wall. Then place a full tile against the wall (Figure 202), overlapping the tile to be cut. Using the measuring tile as a straightedge, score the tile and break it. The cut portion away from the edge will fit perfectly into the edge opening. If you install baseboards and a quarter-round shoe mold, the border tile does not have to be super-accurate.

Corner tile is measured the same way, but you have to measure and cut it twice, in two directions. On an outside corner, make a template with paper, shaping it to fit, and transfer this shape to a tile.

Remember: resilient tile should not be placed directly over boards because their seams will show through. If for any reason, however, the tile must go over boards, apply an underlayment of ¼-inch hardboard or plywood or roofing felt, securing the latter with linoleum cement. Don't overlap the edges of the felt; butt them.

Wood parquet tiles come in 6-by-6-, 9-by-9-, and 12-by-12-inch sizes. They may be solid or laminated, have a one-piece surface, or be made up of a series of strips. Some squares are tongued and grooved. Most are prefinished. Border tiles must be sawed. They are installed in the same way as resilient tiles. As in strip flooring, leave a ½-inch gap on all four sides of a room with parquet tiles, to allow for expansion and contraction; without the gaps, serious buckling can occur.

In bathrooms and kitchens, sheet floor covering is better than resilient tile, because tile has many joints where water can enter, causing the tile to fail. Sheet goods are best laid professionally.

Ceramic and other kinds of clay tile, however, are excellent for bathrooms and kitchens, and sometimes for family rooms. Their only disadvantage is that they are hard (glass and china tend to

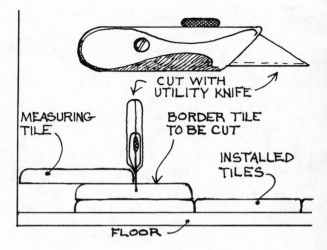

FIGURE 202. *Measuring and cutting a border tile.*

break when dropped on ceramic), and such a floor is not very comfortable to walk on or stand on. Clay tile also has poor sound-absorbing qualities.

Ceramic tile comes in many types and styles, ranging from small mosaics to large pieces of unglazed tile. Glazed tile is preferable because it needs no maintenance other than an occasional swipe with a damp sponge. Glazed tile does not have to be shiny.

Most tile is applied with an adhesive. A so-called mud job is ceramic tile set in mortar. This adds a lot of weight, but if the floor is strong enough under the tile, or is made of concrete, it is the best floor you can have. Tile can also be set on concrete or wood with a quick-set grout—not quite a mortar but certainly not the traditional tile adhesive. It is also good to use.

Ceramic tile, whether individual pieces or small mosaics set in 12-inch-square sizes secured with a paper or mesh backing, is laid out the same way as resilient tile.

Once clay tile is laid, a grout must be applied to the joints. This is usually gray, of the consistency of mortar, and is applied with a squeegee, then packed in tightly with a wood dowel or any piece of wood or a tool that allows you to press the grout into the joints firmly and completely. If this is not done, the grout will fail and you will have to remove it and apply new.

Once the grout is all in, wipe with a damp cloth, then wipe with a dry cloth. When the grout film dries on the face of the tile, wipe and polish with a dry cloth. To make the grout water-resistant (and therefore mildew-resistant), apply clear silicone tile-sealer, and polish with a dry cloth.

Because different woods, resilient tile, and clay tile are of different thicknesses, it is important to adjust the thickness of the underlayment from room to room to make sure the final surface of each type of flooring is level with its neighbors. This is to prevent different levels in different rooms. If you must have a different level in one room, you can put down a wood threshold or an aluminum bead to make the transition. This also applies to any slight differences in floor level between addition and existing house.

chapter 17

The Penultimate

Indoor Trim

It's time for the finishing touches on the inside. That means interior trim and woodwork. Then all you have to do is apply exterior stain and/or paint, varnish, or wallpaper.

You can install plain or fancy woodwork as window and door trim (casings), baseboards, and ceiling moldings (Figure 203). There is Colonial casing and base, belly casing (a little fancier than Colonial), modern (clamshell) molding and casing, and plain 1 x 3 or 1 x 4 or even 1 x 5 boards to use as molding. The belly casing, sometimes hard to find, is Colonial in style but commonly used in Victorian design. All these trims range in width from 1½ inches to 4½ inches and wider. All you have to do is choose a style and install it. The style may depend on what the existing house is done in. The addition trim can match or be different from the existing trim.

The door and window casings are the only parts of a setup window or door that are not provided. Extra-wide jambs necessary for an extra-thick wall in a superinsulated building also are not included. Their installation is described in Chapter 10. Before installing a window casing be sure the jamb is even with the interior wall finish (Figure 204). Once this is done, the rest is easy.

The first trim to go on the interior side of the window is the stool (inside sill), which is notched to fit over the sill and butt up against the bottom

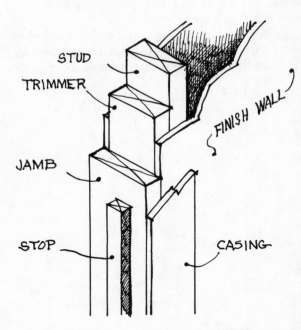

FIGURE 204. *The window jamb must be flush with the inside wall finish in order for the casing to fit properly.*

144

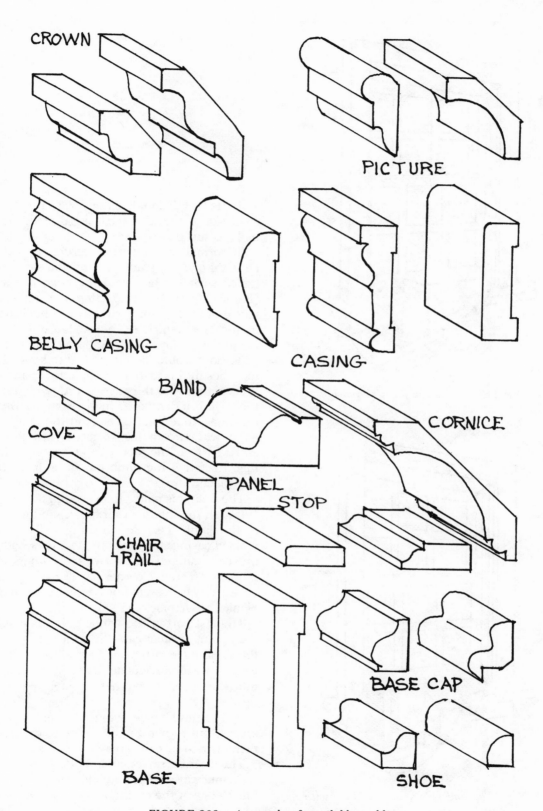

FIGURE 203. *A sample of available molding.*

The Penultimate

FIGURE 205. *Casing is flush to the jamb.*

FIGURE 206. *Top and side stops are flush with the casing or indented.*

sash. Its length should extend ¾ to 1 inch beyond each side of the casing. Use 7d or 8d finishing nails to install.

In normal or standard construction, with regular 2 x 4 or 2 x 6 studs, the top and side casing is nailed so it is flush with the edge of the jamb (Figure 205). Then a top stop and side stops are nailed onto the jambs so they are flush with the casing or indented ⅛ inch (Figure 206).

But with a super-wide wall, the jambs are extra wide, so it would be impractical to build side and top stops in this fashion. Install the casing so that it covers most of the edge of the jamb, but leave the casing indented ⅛ inch (Figure 207). Then the stops, much narrower than the jamb, can sit on it without coming near the casing (Figure 208). In fact, the stop can cover any joint that might occur if the jamb has to be in more than one piece.

The casing is nailed to the rough opening studs (through the interior wall finish) and to the edge of the jamb. Use 8d finishing nails. Milled or molded casing, with carving on one or both sides, must be mitered at the two top corners (Figure 209), or decorative corner blocks can be installed. These may be hard to find. If they are, you can use blank corner blocks (Figure 210).

The last part of the window that is installed is the apron, or stool cap, the trim that goes under the stool. Its length should equal the width of the casing outside to outside. Drive 8d finishing nails into the apron through the stool, and into the header.

Awning and hopper windows usually don't have a stool, but rather another piece of casing. Casement windows can have either treatment.

The interior trim on exterior doors is handled in the same way as that for windows, except that the inner edge of the casing is indented ⅛ to ¼ inch from the edge of the jamb to give it a finished look

Indoor Trim

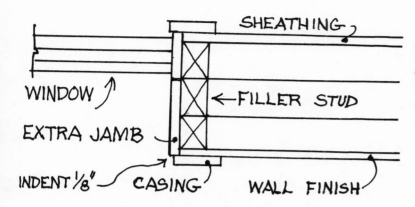

FIGURE 207. *An extra jamb covers the super-thick wall, because a stop will not come out that far. The casing is indented ⅛ inch.*

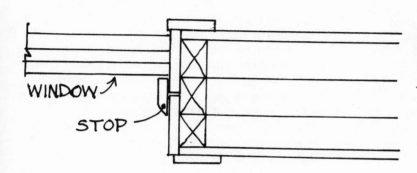

FIGURE 208. *A stop covers part of the window jamb; it also covers the joint between the two parts of the extra-wide jamb.*

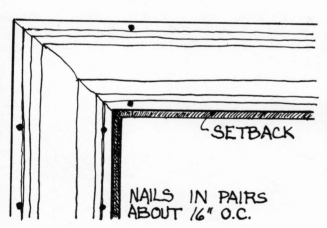

FIGURE 209. *Milled or molded casing must be mitered at the corners.*

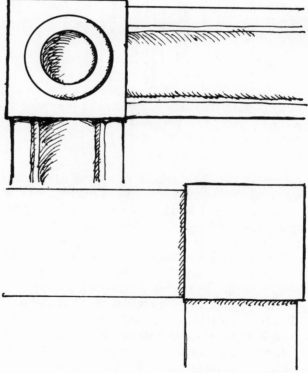

FIGURE 210. *Another way to handle milled or molded casing is to use corner blocks, plain or with a rosette or bull's eye.*

147

The Penultimate

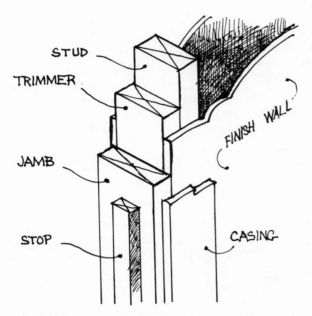

FIGURE 211. *The stop on the exterior door jamb is generally built in, a part of the jamb itself. Casing therefore is indented from the edge of the jamb.*

(Figure 211). The bigger the indentation, the more detail you will get. You can install it by eye or use a strip of wood or hardboard of the right thickness to get the right indentation and to keep it consistent.

Nail with 8d finishing nails through the thickest part of the casing (if it is tapered); use 5d or 6d nails through the narrow part, which goes next to the jamb. If you use a regular board of uniform thickness for the casing, use 8d finishing nails. Nail in pairs and keep them 16 inches apart.

Interior doors are best bought set up; that is, with jambs and one side of the casing intact. The other side of the casing is not on the setup door because the unit has to be set in the rough opening (Figure 212). If you don't use or cannot get setup doors, the work must be very precise so the door will fit, open, and close properly. Sometimes you can buy custom jambs, without the door or casing.

If jambs, casing, and door are installed in place, here is where dimensions must be precise. Jambs must be installed so the clearance between door and jamb is ⅛ inch at the top, ⅛ inch at the latch side, and 1/16 inch at the hinge side (Figure 213). The space between stop and door on the hinge side must be 1/32 inch. Floor clearance is ¼ to ½ inch; more is necessary for a carpeted floor.

The knob goes 36 inches up from the finish floor and the hinges are placed so that the bottom of the lower hinge is 11 inches from the door bottom and the top of the upper hinge is 7 inches from the door top. Exterior doors have the stops as part of the jamb itself. Interior doors usually have a stop that is nailed onto the jamb.

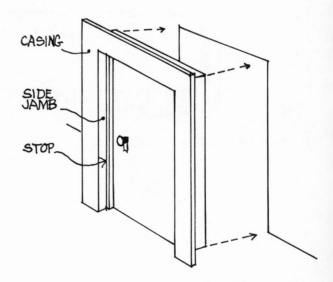

FIGURE 212. *Setup door unit is complete except for one set of side and head casings.*

Indoor Trim

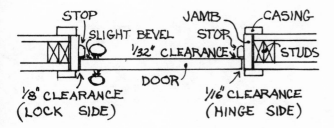

FIGURE 213. *Clearances for a door.*

Fitting the door's hinges can be tricky, too. Interior doors have 3½-inch hinges with loose pins. Exterior doors have two 4-inch hinges, with a third in the center, or lined up with the middle rail if the door is paneled.

Remove the pin from the first hinge and lay half the hinge (one leaf) in position on the edge of the door. The gap between hinge and edge should be at least 3/16 inch. The hinges will be mortised (embedded) into the door edge and the casing so that their surfaces are flush with the wood surface. If you have a router, you can make a precision mortise. If not, having positioned the half hinge of the door edge and made sure it is square, cut around it with a utility knife (Figure 214). A pencil line won't work. To hold the hinge in position while you are cutting, mark the screw holes, drill them, and drive the screws home. Then you can cut hard and deep with the knife, making a perfect fit.

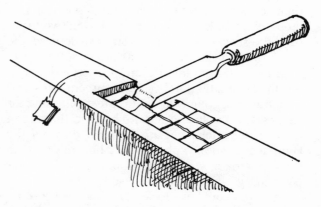

FIGURE 215. *Chiseling of the mortise is made easier by scoring the space with a utility knife in a checkerboard pattern. This avoids splintering the wood. Make sure the knife cuts are all the same depth; the right depth is the thickness of the hinge going into the mortise.*

FIGURE 214. *To make sure a hinge sits tightly on its mortise, install one leaf of the hinge on the surface of the door edge; make a deep cut with a utility knife around the leaf; remove the leaf and chisel out the mortise.*

When this is done, remove the hinge and cut out the mortise with a sharp chisel. Sharpness is important; otherwise you can gouge the wood where you don't want to. To make your chiseling easier, make several cuts in the wood that will be removed, lengthwise and crosswise (Figure 215), with the utility knife. This will prevent your going too deep with the chisel and will make a smoother, better job with a minimum number of fancy tools. Smooth out the mortise so that the hinge fits fully, without rocking.

Now position the door in place in the doorway with shims on the bottom to make the correct gaps at bottom and top. Nail the stops into position temporarily for vertical alignment. Then mark the position of the other leaf of the hinge on the casing. You can connect the two leaves and insert the pin, position the leaf, mark and drill the holes, then drive the screws home. The door won't fit properly, of course, because you haven't yet mortised the second leaf. When you have done so, the door will fit perfectly. At least, it should.

As for locks and latches, most interior doors do not have lockable latches. A bathroom does and sometimes bedroom doors do. Most interior door latches are simple and require holes in the face of the door for knobs and a hole in the edge for the latch, lock, and face plate, plus a hole in the jamb for the striker. Instructions and templates for positioning come with latch sets.

Exterior doors have locks, too. Make sure yours also has a dead bolt; that is, a rectangular or circular barrel that fits securely in a striker opening. Anything less than this, such as a wedge latch that can be set as a night lock, can be jimmied very readily and very quickly. Locks are made for honest people, but you might as well be as secure as possible.

Another trim is base molding (Figure 216). Baseboards range from 1 x 4 to as much as 12 inches in old houses. They can be molded, carved, or plain boards. A plain board is sort of naked, and is best trimmed at the top with a band molding.

Nail the baseboard through the wall into each stud. Use 8d or 10d finishing nails, a pair to each stud. Smaller nails can be used to fasten the band molding, which also should be nailed into the studs. Plain boards can be butted on inside corners and mitered on outside corners; molded boards and band molding must be mitered on all corners.

Since the baseboard is usually installed over a finish floor, it will cover the ½-inch gap between wall and border boards. Sometimes the baseboard is installed before the finish floor or underlayment is put down; in that case the gap is between baseboard and floor. Therefore, a shoe mold, which is a quarter-round piece of molding, is

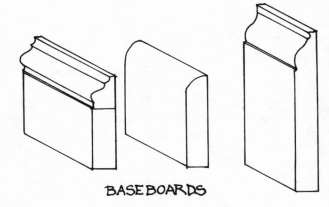

FIGURE 216. *Base molding. At left is molding in two pieces, simply a 1 x 4 (or bigger) board topped by a band molding.*

installed at the bottom of the baseboard (Figure 217). Nail the quarter-round into the baseboard, not into the floor. If the floor boards expand and contract, they will do so under the quarter-round, keeping the gap invisible. If the baseboard covers the gap, a quarter-round is not necessary, but it does give a nice, finished look. A quarter-round is not needed with wall-to-wall carpeting.

Sometimes molding goes at the top of the wall, where it meets the ceiling (called crown molding; Figure 218). It is used to add an extra dimension to the room, or to cover a few sins of omission where wall meets ceiling. It can go next to the ceiling, or in some cases, where the ceiling is extra high, a slightly different kind of molding, such as picture molding, is installed a foot or so below the ceiling.

Because crown molding is generally large and set at an angle, and because most rooms, even newly built ones, do not have square corners, it is

Indoor Trim

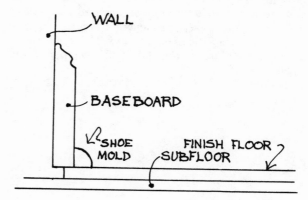

FIGURE 217. *A shoe mold gives a finished look to the joint between baseboard and floor.*

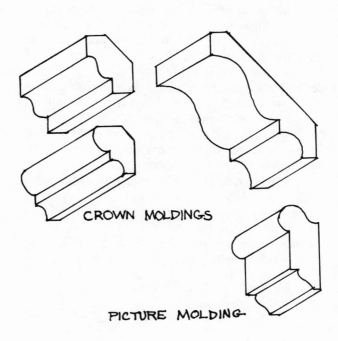

FIGURE 218. *Crown or picture molding is used between wall top and ceiling.*

The Penultimate

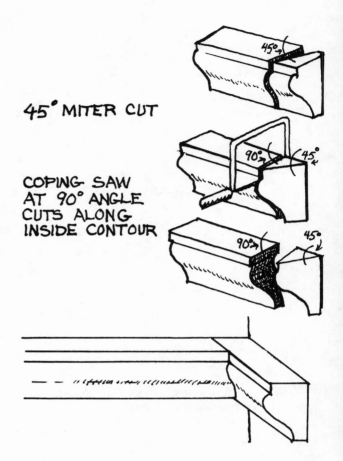

FIGURE 219. *Coping a molding. Miter-cut molding at 45 degrees. With a thin-bladed coping saw, cut end of molding at right angles to its length, following contours formed by the inner edge of the miter. The result will be the same contours in the end of the molding as the molding itself.*

difficult or impossible to miter the corners. So such molding must be coped, which means simply square-cut to match the contours of the molding it butts up against.

It is easier to do this than to describe it (Figure 219). Install one piece of molding so that it butts flush against the corner. Then miter the end of the other molding at a 45-degree angle. Then, with a thin-bladed coping saw, square-cut the end of the molding, following the contours created by the inside of the miter cut.

Relatively new on the market are plastic foam moldings, which are considerably less expensive than wood. They are best used as crown molding and other trims that do not need to stand up against bumping by furniture and scuffing, as baseboards, casing, and chair rails do.

Interior trim is to be stained and varnished, just varnished, or painted, so nails must be countersunk. Don't try to hammer the nails flush with the wood and then countersink them. Rather, stop nailing with the nailhead 1/8 or 1/16 inch from the wood surface; this way you will avoid butchering the wood. Then use a nailset of the right size (buy several) to countersink each nail. Fill with putty and finish (more about finishing in Chapter 18).

If you really want to have a finished look on your woodwork, do what some old-timers do: wherever you plan to drive a nail, cut a small sliver of wood and, if possible, leave one end of it intact; lift up the sliver, drive and countersink the nail, and replace the sliver and glue it down. It has been done, and in fact there are tools for just such sliver-making.

chapter 18

The Ultimate

Paint, Stain, Varnish and Other Good Things, Inside and Out

If wood is kept dry or is protected, it will last indefinitely. The trouble with many houses and buildings is that their wood components will begin to decay if they're unpainted or the paint or stain is allowed to wear or peel off, or the wood is left in contact with the ground or exposed to water.

So to keep your addition and your house free of decay, cover, protect, and keep wood dry. To do this is easy. Beams and posts within are often protected by outer boards, such as sheathing and siding (shingles and clapboards). The protective materials—siding and trim—are painted or stained.

PAINT

In the good (or perhaps the bad) old days, paint was a fine material, able to protect wood against the elements and also resistant to peeling and wearing. But it was full of lead. This was the thing that made it element-resistant, but also made it poisonous to anyone who ingested it. So lead has been banned in paint. Oil paint now contains such metals as titanium, which can't hold a candle to lead in protective capability. And even today oil paints are banned in California because of the toxic materials they give off even after they have cured. Among the substitutes for oil paints developed since World War II are latexes, touted as the be-alls and end-alls. They have not lived up to their potential.

Paint can also fail because of moisture from the inside. Water vapor in the house penetrates the wall finishes, goes through the wall cavities, or through the insulation if there is any, then through sheathing and siding, and out to the paint. Being resistant to water penetration, the paint will allow the water vapor to build up behind it. When the vapor condenses into water, it creates a blister, and when the blister breaks, you get peeling paint. Sometimes if you cut a blister, it will release water, just as the blister on your finger does if you prick it. Latex paint is designed to breathe, to allow the water vapor to pass through it harmlessly. That is good in theory, but it doesn't always seem to work in practice, and the situation is worse when latex is mixed with oil paint. When latex is applied over oil or vice versa, the two different kinds of paint tear each other apart because they expand and contract at different speeds and distances. This can happen even when you have a good vapor barrier inside.

Moisture can also attack from the outside. If it gets under shingles and clapboards, it can raise

The Ultimate

havoc just as if it came to the outside finish from inside the house. So if you really want to paint your siding and trim, do so at some peril.

Painting is not just a matter of slapping a coat of paint on the wood. It is a *system,* which includes a primer coat and two coats of finish paint.

Choose oil or latex, not both. With oil, use a nonporous oil-based exterior primer. Then apply two coats of top or finish house paint. Apply the first coat within two weeks of putting on the primer and the second within two weeks of the first. If you opt for latex, use a primer that is designed to go with latex finish paints. It may be latex or oil.

Do not paint where the sun is warming the siding. Follow the sun around the addition, always painting in the shade. Don't paint when the temperature is under 55 degrees or will go down to 55 during the drying and curing time for the paint. Avoid painting when excessive moisture is present — i.e., right after a downpour.

These rules of painting also apply to the trim: one coat of primer and two coats of trim paint, oil or latex. Some trim, such as setup window and door casing, is preprimed. All it needs is two coats of finish paint.

If the trim has any knots, they must be sealed with fresh, white (clear) shellac before you prime. Window and door casing won't have knots, and neither should the trim that you chose and put on the addition; it should be clear pine. All nails on the trim should be countersunk and filled with glazing compound or exterior putty.

EXTERIOR STAIN

Stain is really the wave of the future. It is designed to go onto bare wood and penetrate it. Since it does not form a skin, as paint does, to protect the wood, it won't peel or blister. It protects by penetration and by making the wood resistant to water. While paint ideally will last for five to seven years, the most you can expect from stain is five years, after which it will show signs of wearing away or fading. But usually only one coat is required to renew it, and there is no scraping or priming.

White cedar shingles, redwood, and cypress will resist decay and are woods that don't have to be stained, painted, or treated with preservative. The white cedar shingles will weather to a silver color. Eventually they will erode away — at the rate of ¼ inch every hundred years or so. Redwood is expensive, alas, and cypress is difficult to obtain. Shingles and clapboards of red cedar also resist decay, but normally are stained.

Stains come in two types — oil and latex — and numerous designations — semitransparent, solid, and preservative types both semitransparent and solid. Semitransparent stains come in several colors; solid stains in virtually every color, as many as paints do. Latex solid stain, says one manufacturer, can be applied over old paint. The preservative-containing stains are useful in making the siding and trim water-resistant but are not necessary if the wood is cedar.

Apply one or two coats of stain; two coats are generally required for solid stain. The timing and temperature rules of painting also apply to staining.

Trim can be stained the same color as the siding or can be a contrasting color. For light colors on trim — indeed, for trim generally — use solid color stains. Window sashes and sills should be painted.

Accessories such as door and shutters are best painted with an oil-based primer and acrylic latex trim paint, although solid stain should work, too. On doors, make sure that all edges (tops, bottoms, and sides, too) are painted, as well as the outside face. On the inside of the door, use an undercoat and one or two coats of latex interior paint. If you like the idea of a stained or natural door, stain will do well. Avoid varnish on an outside finish. It will darken, turn yellow, and peel within a year or so, and if it does peel and the bare wood weathers to a gray, you must sand down to new wood before revarnishing. Varnish on a weathered wood surface will turn black immediately.

Exterior sills are best stained with a semitransparent one. Decks also should be stained with a semitransparent preservative stain.

INSIDE WORK

Painting, staining, and varnishing interior trim is easy because there is no need to protect it from the weather, only from wear and tear and to make it look good. Trim does not have to be protected from interior moisture, but doors do because they will swell when they absorb moisture and thus will be difficult to close since they won't fit well in their frames. So make sure all edges, front, and back are primed and painted or varnished, with or without stain.

A good paint job should stand up well even in rooms where extra moisture is generated, such as bathrooms and kitchens. Use a good latex enamel undercoater and latex or oil top coats. Also, moisture in these rooms can be kept to a minimum with proper ventilation, such as an exhaust fan.

The real sufferers in bathrooms are the painted metal covers on baseboard heating units. Moisture affects the paint on them more than it does the paint on wood surfaces. This is particularly true when there are a lot of males in the family. Not intended to be funny; it's a fact of life. There is no cure; on metal that has peeling paint and rust, sand off the peeling paint and the rust, apply a metal primer and one or two coats of an oil-based indoor-outdoor enamel — and wait for your sons to grow up and move out.

As for interior windowsills, they should pose no peeling problem unless there is a great deal of moisture condensing on the window glass and dripping down to the sill. It has happened, and it can be serious. Proper glazing (prime windows, perhaps double-glazed, plus storm windows and inside storms) can prevent excess condensation. So can good ventilation in the area, to prevent buildup of water vapor.

Is there a particular order in finishing a room? Yes: ceiling, trim, walls, floor. The ceiling is first because it is isolated, and also it is easy to protect the floor from any drippings, or the drippings don't matter because the floor is going to be done anyway. Trim is second because it is easy to paint the edges next to the as-yet unfinished walls; if you slop anything on the walls it will be covered by the wall paint or paper. Painting walls is easier to do against newly painted trim. Wall painting or papering comes third; and floors are last, because everything else is done.

Ceilings take two coats of white latex ceiling paint. You can use some color, but remember that making a ceiling darker than the walls will lower it visually. Sometimes a ceiling is scrolled, by using a texture paint (Figure 220), or given a rough finish by spraying a plasterlike material on it. Generally, however, it is best to leave the ceiling smooth and just painted.

Before you paint interior trim be sure all nails are countersunk and filled with a wood putty of the type you mix with water. It will dry quickly and can be smoothed off with a wet sponge or sanded.

Latex paint is king for interior-trim paint, even in bathrooms and kitchens. Apply one coat of latex enamel undercoater; this is absolutely essential

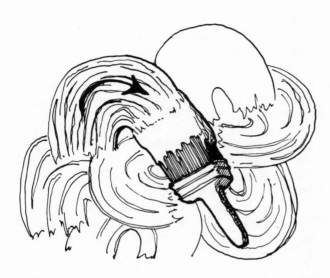

FIGURE 220. *A scrolled ceiling is made with texture paint and a stiff, dry brush.*

(on new wood as well as over old paint) to resist chipping, particularly on kitchen cabinets. Then apply one or two coats of latex eggshell, semigloss, or gloss enamel. For trim, semigloss is good, but the relatively new eggshell finish—neither flat nor shiny—is good, both for trim and for walls. If the finish coat is close in color to the white undercoater, you may need only one coat. If the color is slightly or somewhat lighter or darker, you will need two. Don't try to make one coat cover; chances are very good that it won't. Sometimes, however, you can tint the undercoater with universal tinting colors to reduce the need for a second finish coat.

If the wood has knots, cover them with fresh, white (clear) shellac or white pigmented shellac before putting on the undercoat. Two coats of shellac are better than one, and even then there is no guarantee that they will keep the knots from bleeding through undercoater and finish coats. The older the wood, the less likely that the knots will bleed.

Trim can also be varnished, but varnish applied directly on most softwoods will come out very yellow. It is best to stain first and then varnish. Use a dye stain or penetrating oil stain that is as transparent as possible. Avoid heavily pigmented stains; they look like paint. Oil stains are better than latex ones.

Apply stain with a brush or cloth. You can control the color by wiping the stain off with a cloth. The sooner you wipe, the lighter the stain; the later, the darker. Two coats will also darken the stain. After staining apply two coats of semigloss, satin, or flat varnish.

Filling holes with putty, by the way, will not work with a dye stain, because the stain will not cover the light-colored putty. You can buy a colored putty or make your own by mixing the putty powder with latex stain. You may have to experiment to get the right color, but follow this rule of thumb: a putty that is darker than the final finish will be much less obtrusive than a lighter putty. And remember this: the dark putty is likely to stain the wood as you apply it. It is difficult to remove, particularly from an already stained surface. So before puttying, apply the stain and one coat of varnish. Then apply the colored putty. Because it is water-soluble, you can wash off any smears and smooth off the patching putty at the same time. Then finish with your final coat of varnish.

For walls, two coats of latex wall paint will look good and stand up to wear and tear and steady washing. Like ceiling paint, the first coat will prime and seal, the second coat will finish. For both ceiling and wall paint, do not try to cover with just one coat; the result could be too thick a coat and a poor job. Two thin coats are always better than one thick one.

Wood floors require a high-gloss urethane varnish. If you can find water-cured urethane, all the better: it may stay a little sticky in dry weather and it dries best in humid weather, but once it dries and cures, it is very wear-resistant. The high-gloss varnish is more resistant to wear than a semigloss. If you like a semigloss finish on a wood floor, give it two coats of high-gloss and a third of semigloss. Generally two coats are enough for a hardwood floor; three are good on a fir or pine floor.

When using urethane varnish, be sure to follow label instructions. When the instructions say to put on the second coat not less than four hours after the first but not more than twenty-four, or eight, or what have you, they mean it. If the first one or two coats get too dry and hard, the succeeding coats will peel off in sheets, unless the previous coat is hand-sanded to remove or reduce gloss and to roughen the finish.

Wallpaper

Suppose you want wallpaper on the walls instead of paint. Even though it's more expensive than paint, it can be elegant, and it is satisfying to install. Once it's up, you are virtually finished with your project.

If your walls are plastered, or even skimcoated with plaster, all you have to do is apply a glue size and put up the paper. Glue size is a powder that you mix with water and put on the walls like paint. It's a funny material: it makes it easier to move the paper when you're trying to position it during installation; it makes the paper stick better; and it makes it easier to remove the paper when that time comes or if you goof.

If your walls are plasterboard, however, this is another matter. Plasterboard has a paper face of its own, and when wallpaper is put on it, it's the very devil to remove without removing the plasterboard face paper, and ruining the whole wall. So the plasterboard must be sealed to make it water-resistant. Apply one coat of vapor-barrier paint or two coats of flat oil-based wall paint. Use white; it's a neutral color. Then coat with the glue size and put up your paper.

There are many kinds of wallpaper: vinyl-coated, canvas types, flocks, foil-backed, Mylar. Some papers are strippable, which means they can be removed by simply grabbing a corner and pulling. There is prepasted paper, which needs only soaking in water to activate the paste. These papers are good, but care must be taken not to soak them too long and thus wash off all the paste. It is a good idea to hang prepasted papers with paste anyway. This is done in the normal manner.

Wallpaper comes in double and triple rolls. A single roll is 32 square feet, a good number to remember when you estimate how many rolls you need for a room. To measure the wall area of a room, multiply the height by the width of each wall and add them all together. Ignore windows and doors when you calculate this area.

Sometimes a double or triple roll is not all one piece; that is, its length is interrupted. When you run into a cut roll, you are not being gypped because extra paper is provided, but it's a shock to find two pieces in a roll.

Measure a wall and cut the paper into strips a few inches longer than the distance from ceiling to baseboard. Cut above where you want the pattern to disappear into the ceiling. A utility knife cuts paper well, even when it is wet, but make sure the blade is sharp. Change blades often,

The Ultimate

certainly when the swipe of the knife tears the paper. You will also need a paste brush, paste bucket, a smoothing brush, and a long straight-edge. A little wooden roller, called a seam presser, might also be useful in sealing seams.

You will also need a table to work on. Use a door off its hinges, or a piece of plywood, put over the door or used alone, set on the backs of chairs or laid on a bed. To prevent paste from getting on the table, cover it with newspapers; the ink won't get on the wallpaper.

The paste should be of the type compatible with the paper. Mix it according to instructions. Cellulose paste will seem quite thin; wheat paste is a little thicker, but not much. Both are excellent.

Wallpaper must be hung plumb. If the ceiling, floor, or another wall is not level or plumb, ignore it. To hang the first strip, start in a corner, preferably behind a door (Figure 221), and mark a plumb line on the wall, away from the corner, ½ inch less than the width of the paper strip. Use a plumb line or long level to make the mark. For instance, if the strip is 27 inches wide, make the mark 26½ inches from the corner. This will allow a half-inch of the paper to turn the corner; more than that will result in wrinkling and a mess that you won't be able to correct.

Now put the first strip facedown on the table and apply paste on the bottom half of it first (arrows on the back of the paper sometimes show the direction in which the paper should go); paste down the center of the strip lengthwise, then paste the edges by brushing from the center to the edge in both directions. Then fold the bottom half over itself. Then paste the top half. This is done because most tables are not as long as the length of the strip and also because a folded strip is easier to handle.

Line the strip up with the plumb mark. Let the top edge overlap onto the ceiling. Smooth the

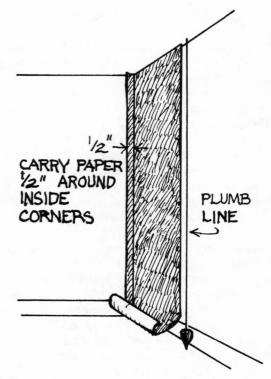

FIGURE 221. *Start the first strip of wallpaper in a corner, preferably near a door. Make a vertical line on the wall ½ inch less than the width of the paper from the corner; this will allow paper to turn the corner by ½ inch. Don't turn any more than this; the paper will wrinkle hopelessly.*

paper down with the smoothing brush, working out bubbles and wrinkles by pushing them toward the edges of the strip. When you are partway down and the paper is sticking properly and positioning accurately, unfold the bottom half and keep going. Get everything as smooth as possible and push out bubbles; small wrinkles will disappear as the paste dries. After all is smoothed and

Paint, Stain, Varnish and Other Good Things, Inside and Out

the position is correct, cut the paper at the ceiling and baseboard.

A cutting trick, particularly at the ceiling, is to place a wide smoothing knife or long straightedge in the corner between wall and ceiling and cut along its edge with the utility knife.

After you have smoothed and cut, go over the strip with a damp sponge to remove excess paste from the paper, ceiling, and baseboard. Keep a bucket of water handy for the sponge and change the water often.

Sometimes the paper will have an extra-long drop in its pattern, which means you can waste a lot of it when you line up the pattern of the next strip to match the pattern of the first strip. To avoid this, instead of taking the next strip from the same roll, take it from a separate roll. When you match the pattern, do so at eye level; it may be a bit off at the bottom, but no one, including you, will notice. If the pattern doesn't seem to match no matter how hard you try, take it back to the store for replacement.

To go around door and window casings, measure and cut the strip to fit the space between the previous strip and the casing. Then use the paper you cut off to fit over and under the window casing. Do the same when you come to the second corner: if the space you need to fill to a corner is only 10 inches wide, cut a strip 10½ inches wide so that it will turn the corner by ½ inch. Then use the piece you cut off to continue on the second wall. Draw another plumb line on the second wall because if the walls are not quite perpendicular to each other the second strip may not be plumb. You can also overlap the ½ inch of paper that turned the corner because that ½ inch may not be plumb. Again, no one will notice.

If you get an unremovable wrinkle, which may show only after the paper has dried, cut or tear the paper along the wrinkle and apply hot water to soften the paste. Then smooth it out. A torn edge is less noticeable than a cut one. And if the cut or tear is high on the wall, lap the bottom edge of it over the top edge so that you won't see the joint when you look up. If the cut is toward the bottom on the strip, lap the top edge over the bottom edge. There are lots of ways to fool the eye.

You're all done except for the shouting!

chapter 19

Before and After

Converting Garages, Breezeways, and Basements

In the preceding chapters we've described the building of an addition that is essentially a new structure, attached to the existing house. Another good kind of addition is made by converting already built breezeways and garages to sleeping or living quarters, or family rooms — the use of the new space depending, perhaps, on how far away it is from the house (Figure 222). If there is no covered breezeway connecting garage and house, one can easily be built. Still another great and sometimes untapped source of expansion is the basement.

Garages

A double garage is about 20 by 20 feet or larger; a single garage is about 12 by 20 feet. A breezeway may be narrow (6 feet or so) and as long as the distance between house and garage. Converting garage and breezeway can add as little as 260 square feet or as much as 500 square feet to living space. A breezeway can be widened, but this requires rebuilding it. If the garage roof is high or steep enough, space under it can be made into a full second floor or a sleeping loft or other type of hideaway, or, at the least, into storage space. The ceiling can follow the contour of the roof for an automatic cathedral ceiling.

First you must determine if the concrete slab in the garage is on a real foundation or is only a slab on the ground. The older the garage, the more likely that it is simply a slab. If the garage is attached to the house or is part of the house or is in the basement, the more likely it is to have a full foundation.

Without a foundation around the perimeter of the slab, the garage will tend to rise and fall with freezing and thawing of the earth underneath it in the winter. If you were to convert it to living space, it would be hazardous to install plumbing. Also, a connector built on a foundation of its own would have to be independent of the garage, that is, not hooked up to it, because the firmly anchored connector wouldn't move, but the garage would, and would tend to tear apart any connector you made. It is like a porch attached to the house by way of roof framing nailed to the house itself but without a foundation; the porch moves, the house does not. Sooner or later you have a drooping porch.

So there really is not much point in converting a garage or breezeway that doesn't have a foundation. There are ways, however, to install one.

New Foundation
To determine whether your garage or breezeway has a foundation, dig around the perimeter on

Converting Garages, Breezeways, and Basements

FIGURE 222. *A two-car garage and a medium to wide breezeway are ideal candidates for conversion to living space.*

any side but the car-door side. If there is concrete or concrete blocks down three or so feet, you can be sure that that is the foundation. If the concrete goes down only a few inches or a foot, there is a foundation of sorts, but it is inadequate in northern climates. A foundation must go down below the frost line, some four feet in extreme northern areas. Check your community's building department, building inspector, or building code.

A garage that has no concrete slab but only a dirt floor can be lifted bodily up several feet, set on timbers, and a proper foundation with footings installed (Figure 223). In this case, there is no point in pouring a concrete floor; instead, floor joists are installed on 4 x 6 wood sills set on the new foundation, a floor built, and the structure eased back into position on the new floor.

If there is a concrete slab, you have some choices. You can raise the wood structure, break up and remove the slab, and build a new foundation. Another way is to break up the slab only around the perimeter and install a foundation and a wood joist floor above the old slab. These choices are expensive because they require a rigger or house movers to raise the structure. If you must do this, you could take advantage of the rigger or house movers and position the garage closer to the house. If the garage structure is sturdy and worth retaining, this might be a way to go. Certainly, beyond the raising and/or moving of a structure, you can do most, if not all, of the conversion work yourself.

A new foundation at the car-door end of the garage is installed the same way as on the other sides, but the driveway—asphalt or concrete—must be cut away so that the trench can be dug for the foundation. You should also remove the garage door, since this will become a wall—one with windows if the door end faces south, or, if you plan to make the old driveway a patio, one with sliding glass doors or, even better, French doors.

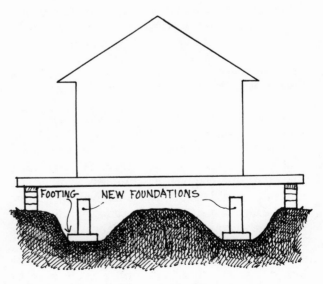

FIGURE 223. *If a garage has no foundation, it must be raised and held in a raised position while the foundation is installed, with footings.*

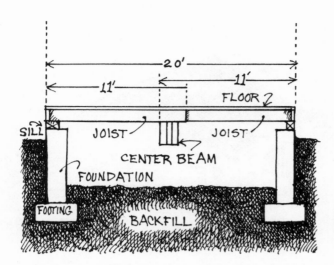

FIGURE 224. *A new foundation on a garage is treated like the crawl space of any construction; joists are applied on sills, a floor put down, and the structure installed on the floor.*

Once you have a new foundation, install a floor on top of it just as you would in building a new addition or building (Figure 224). Place a 4 x 6 sill on the foundation, securing it with anchor bolts. For a garage 10 to 12 feet wide, install 2 x 8 floor joists. For one 20 feet wide, 2 x 8 joists will do, but they must be supported by a central beam. The joists do not have to be the full 20 feet long; they can meet and overlap by 6 to 12 inches on the center beam. Header joists are placed at the joist ends, connecting them, and stringer joists are placed parallel to the joists at each end of the garage.

Usually when a new floor with joists is installed on top of a new foundation, there is a very shallow crawl space under it. Cover this with 6-mil polyethylene to keep moisture from coming through the ground and finding its way into your new floor. Insulation here with a polyethylene vapor barrier will also block the migration of water vapor.

When you set the garage back down on the new foundation, sill, joists, and floor, you will find your structure has risen about 11 inches, so you have a problem with the outside finish (Figure 225). Assuming the clapboards, shingles, or other siding is intact and usable, you still have exposed the new header and stringer joists and the sill, so you will have to cover this gap with sheathing boards and siding to match what already exists.

If the siding is clapboard, the bottom one can be pulled out to see what's under it. It may be another clapboard or an undercourse of shingles, set so that the bottom clapboard flares slightly at the bottom, acting as a drip edge. In this case, remove the shingles or clapboard from under the bottom clapboard. Nail on a piece of plywood to match the thickness of the sheathing boards, or use sheathing boards, and apply over it 15-pound roofing felt. Install an undercourse of shingles along the bottom, over the felt, and then put on new clapboards, at the same exposure as the old, until you reach the new foundation.

If matching the exposure of the new clapboards to that of the old means that the lowest (bottom) clapboard is going to have a very short exposure, it is wise to measure the entire extent of the area where the new siding will go and divide the number of clapboards needed to determine the reduction in exposure necessary. Or you can gradually reduce the exposure as you go down to the sill. That 11 inches you have gained will allow you to put up several courses of clapboard (Figure 225).

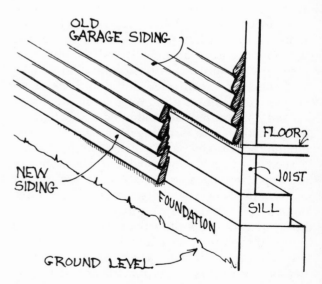

FIGURE 225. *Where new sill, header, and stringer joists are exposed when a new foundation is built, new siding must be installed.*

The same measuring and dividing technique is used if your siding is shingle. You may need a slate puller to remove hidden nails (Figure 226). While you can install clapboards from the top down, you must install shingles from the bottom up. Put on an undercourse along the bottom, making sure it is level, and put a starter course over it. If the gap you are covering is 11 inches (from the bottom of the old siding to the bottom of the sill), and the original shingles are exposed 6 inches, you can install two courses with a 5½-inch exposure. When you get to the second course, you may have to cut a few inches off the tops of the shingles so that they will fit properly when shoved under the original shingle course.

Another technique is not to continue the siding at all, but rather to build a water table, which is a board (1 x 12 in this case) installed just below the original shingles or clapboards (Figure 227). Attach metal flashing under the bottom course of the original siding and bend it so it will form a drip edge over the water table. The water table is painted or stained to match the siding or trim.

Existing Foundation

It is a simple matter to install joists for a wood floor in a garage with an existing foundation and slab. But first you must take care of the driveway area.

Cut away the existing driveway outside the foundation so that the foundation top is at least 6 inches above the ground level. This is to prevent water from coming in under walls or doors. If the driveway slopes down to the car doors, there are two solutions:

1. Build a dry well at the end of the driveway about a foot away from the garage (Figure 228). This will intercept water coming down the driveway. It can be 4 to 6 feet deep and 2 or 3 feet wide, and filled with coarse crushed stone or large

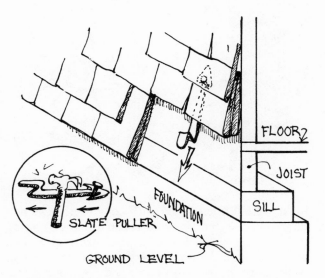

FIGURE 226. *Shingles are a little more complex to remove. A slate puller, used to cut hidden nails, helps. Filler shingles are then installed, and the tops of the last course next to the existing shingles are tucked under the existing shingles.*

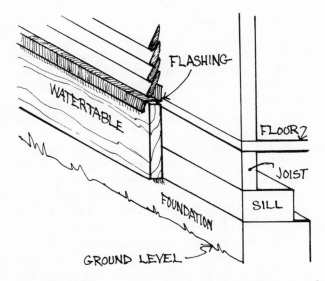

FIGURE 227. *A water table (1 x 12 board or one of dimensions that will fill the gap) is installed under flashing.*

Before and After

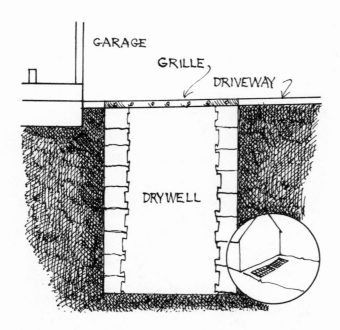

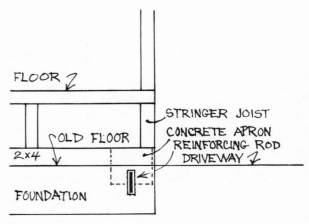

FIGURE 228. *A dry well, with proper grille, is installed over the driveway in front of a converted garage, to prevent water flowing down the driveway from entering garage.*

FIGURE 229. *A concrete apron is at the same height as the bottom of the stringer joist. It covers the area where garage doors once did their duty.*

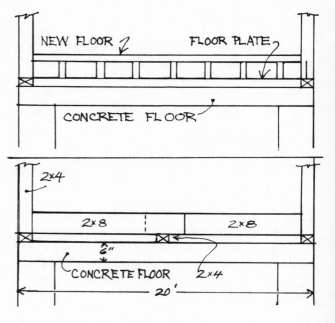

FIGURE 230. *Two views of floor joists that sit on the 2 x 4 floor plate of the garage wall. This technique is used when the garage already has a proper foundation, and floor boards are needed above the concrete floor.*

rocks, or left open, lined with concrete blocks on their sides, so that water can seep into the ground through the sides. Install a steel grille over the opening. Or, if the area is earth and grass, the dry well can be topped with earth and seeded over. It will still intercept any downward-flowing water, but not as efficiently as the open one with a grille would.

2. Build a concrete apron or ridge along the edge of the garage so that it sits 6 inches above the driveway or ground, or is at the same height as the tops of the interior floor joists (Figure 229). To secure it to the existing concrete, drill holes in the concrete and insert steel reinforcing rods 4 inches into the old concrete, sticking up about 6 inches. Space the rods every 2 feet. Build forms, coat the concrete floor inside the forms with roofing cement, and pour the concrete. The roofing cement will form a bond that should keep out water.

As to the floor: the floor joists sit on the original sill of the garage, which is the bottom plate of the walls. It is a 2 x 4, so the joists sit 1½ inches above the concrete slab (Figure 230). In a single garage, 2 x 8s or 2 x 10s will serve. They will in a double garage, too, but—as in the floor on a new foundation—they must be supported by a center beam. Make sure the joists are level along their lengths and at right angles to their lengths. Set them in place and toenail them to the plate. If they are on 16-inch centers, nail them to the sides of the studs. If you set them on 24-inch centers,

164

toenail them to the bottom plate only. To keep them steady, build 2 x 4 fire stops along the tops of the joists and set them between each set of studs (Figure 231).

There is some disagreement about where vapor barriers should go when a floor is set above a concrete slab. I suggest laying 6-mil polyethylene on the concrete to prevent moisture from coming up from the ground through the concrete and into the floor. Then you fill the floor cavity with fiberglass insulation: 7½ inches for the height of the joists, plus 1½ inches for the thickness of the floor plate equals 9 inches, so you can lay one layer of 6-inch unbacked fiberglass batts first and another layer of 3½-inch unbacked fiberglass over this. If the joists are 2 x 10s, add to the insulation appropriately. Then cover the tops of the floor joists with 6-mil polyethylene as a vapor barrier. Now these two layers of polyethylene make a double vapor barrier, which is usually taboo, but in this case it is workable because there is no air space for water vapor to get into the cavity between the two vapor barriers.

Install ⅝-inch plywood on top of the joists as a subfloor. Later you can put oak strip flooring or plywood underlayment on it, and top with resilient tiles, ceramic tiles, or carpeting. You're ready for the walls!

Walls and Ceilings

Many garages have simple stud-and-sheathing walls, with no inner finishes. So with those it is easy to make the conversion a superinsulated one. Build a second wall of 2 x 6 studs just inside the present wall, separating them by about an inch, or one of 2 x 4s set 2 inches from the existing wall. Locate the studs so that they do not line up with the existing ones. Fill the existing wall with 3½-inch unbacked fiberglass batts or rolls, and insert 6-inch unbacked fiberglass in the new wall and

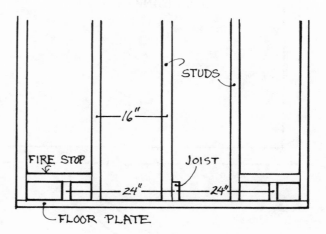

FIGURE 231. Fire stops are installed between studs to steady joists that do not butt up against the studs. Joists that do butt are nailed to the studs.

put 6-mil polyethylene on the face of the studs.

If your garage is plastered on the inside or has any other kind of finished interior wall, you can have insulation blown into the cavity and can add an interior, insulated wall for superinsulation. Taking off an existing inside wall surface and installing batt or roll fiberglass insulation is more work but will cost less. For concrete or concrete-block walls, the only way to achieve superinsulation, or something close to it, is to build a 2 x 6 stud wall inside and fill it with fiberglass insulation.

The garage will have ceiling joists—maybe just a few. If that is so, they are there not so much for a ceiling or second floor as just to keep the roof from spreading the walls (Figure 232). The interior wall with its single top plate is nailed to these joists. Window and door treatment, by the way, is the same as that in any other addition, with adjustments for the different thicknesses of the

Before and After

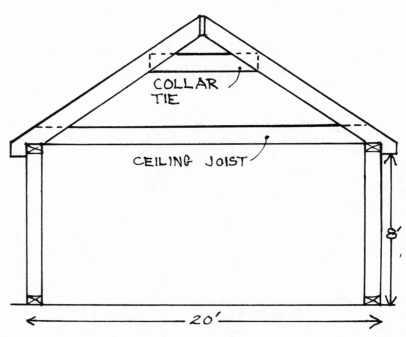

FIGURE 232. *A garage may have a few, or several, ceiling joists, even if there is no ceiling applied to them. The collar ties, probably even less numerous than the ceiling joists, keep the roof rafters from spreading. Therefore, they should remain if ceiling joists are removed.*

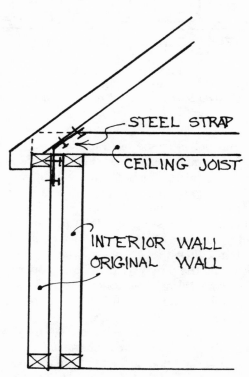

FIGURE 233. *Metal straps extend from the outside wall to rafters to keep them from spreading the garage walls. The inside wall extends from ceiling joists to floor.*

walls. If a window already exists in the garage, you must allow for it on your interior wall.

You may want to leave the ceiling joists to support a ceiling and a second floor. In that case, they should be set on 24-inch centers. But they can be removed if you want to use the garage roof as a cathedral ceiling. First, keep collar ties high in the ceiling, connecting the rafters (2 to 4 feet below the peak), or install some if they are not there (Figure 232). Nail metal connectors from the rafters to the inside of the outside wall (Figure 233).

If the ceiling joists are removed, the new inner wall will be attached in a slightly different way so that it will act as a tie to the rafters. Nail a top plate to the rafters at the location of the top of the wall (Figure 234) and cut studs to fit to the top plate and a bottom plate. With the metal connectors, this will make a good bond between wall and rafters, and, with the collar ties, will prevent spreading of the walls or sagging of the rafters.

When rafters are 2 x 8s or so, there is not much point in exposing them as beams. So fill them with 6 inches of unbacked fiberglass insulation. This will allow an air space between insulation and roof boards. To make that air move, install soffit vents on the underside of the eaves and a ridge vent (Figure 235).

But a mere 6 inches of insulation isn't very

Converting Garages, Breezeways, and Basements

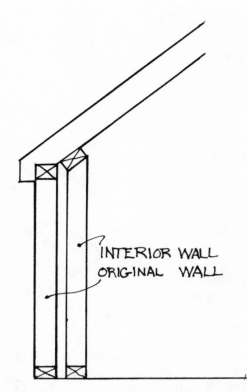

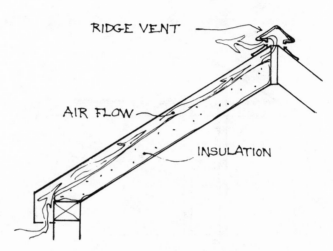

FIGURE 234. *How the interior wall is set against a top plate nailed to the rafters. This technique is necessary when ceiling joists are removed. Collar ties must stay in place.*

FIGURE 235. *A cathedral ceiling must be well insulated, but there also must be an air flow between the insulation and the roof sheathing. This is how to do it.*

much; in fact, not superinsulation at all. You can increase the amount by nailing 2 x 2s to the bottoms of the rafters (Figure 236) to extend their depth (if the rafters are 2 x 6s, extending them is a must). Then nail 1 or 2 inches of Styrofoam or High R Sheathing to the rafter bottoms, cover with 6-mil polyethylene, and put up your ceiling finish, using extra-long nails to go through the insulation and into the rafter extensions. Depending upon the type and thickness of the rigid insulation, you will attain an R factor of 27 to 35 — fair to middling in regard to superinsulation. Incidentally, if you apply High R Sheathing with a heavy foil, tape joints with duct tape and you won't have to put up the polyethylene.

If you don't plan a cathedral ceiling, install ceiling joists at 16- or 24-inch centers and staple 6-mil polyethylene to their bottoms. Use 2 x 8 joists on a single garage and 2 x 12s on a 20-foot-wide one. Put up your ceiling finish and then install two layers of 6-inch unbacked fiberglass above it, leaving the eaves open and ventilating the space above the insulation. This will give you a superinsulated ceiling.

Some garages have attics that can be used as a second floor or loft; if the roof is steep enough, you can make a second floor or loft. A full second floor requires a staircase; in this case a spiral staircase looks good and saves space. A loft, taking up one

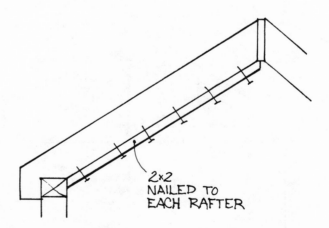

FIGURE 236. *If the rafters are not deep enough to accommodate at least 6 inches of insulation, nail a 2 x 2 extender to the edge of each rafter. Then, when 6-inch insulation is installed, there will be plenty of air space between it and the roof sheathing.*

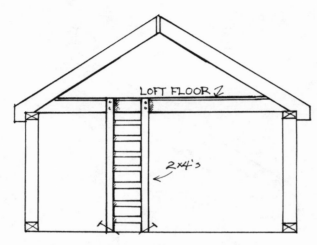

FIGURE 237. *The easiest way to reach a loft is by a permanent, vertical ladder.*

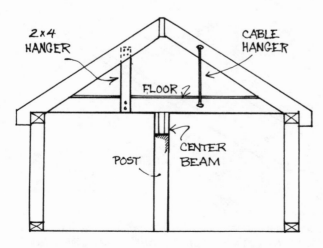

FIGURE 238. *Three ways to support ceiling joists so they can act as floor joists.*

third to one half the space, is easily reached with a permanent ladderlike staircase (Figure 237). In a loft arrangement the rafters should be insulated as they are in a cathedral ceiling.

To build a loft, 2 x 6 fir decking is good; it acts both as flooring and as ceiling finish. It is put over joists set at 24- or even 48-inch centers. It is a good idea to put a center 6 x 10 beam midway between the joists of the loft floor. The beam can be supported by posts below or hung by cables or even 2 x 4s from above (Figure 238). The loft should have a railing, at least, and provision for draperies or other means of obtaining privacy. The railing should be 30 to 36 inches high and can be made of regular rails and balusters, or set with posts bolted into the joists and railings and balusters put between them (Figure 239).

Under any circumstances, with cathedral ceiling, second-floor living space, or a loft, the gable-end walls of the garage must be insulated. An extra wall should be installed, insulated with 6 inches of unbacked fiberglass and a polyethylene vapor barrier. The existing walls at each gable end should be insulated with 3½-inch unbacked fiberglass.

A breezeway is converted the same way as is a garage. If it is narrow, leading to a double garage, there may be no need to expand it. If it leads to a single garage, widening it would create an *L*-shaped garage-cum-breezeway addition (Figure 240). But widening a breezeway means a new foundation and a roof; it requires rebuilding the entire breezeway, and that is no longer a mere conversion.

The heating of a converted garage can be planned with an eye to solar sources. If a gable end faces south, windows can be installed there (Figure 241). This is particularly appropriate for the wall that replaces the garage door. Or skylights can be put in a part of the roof that faces

Converting Garages, Breezeways, and Basements

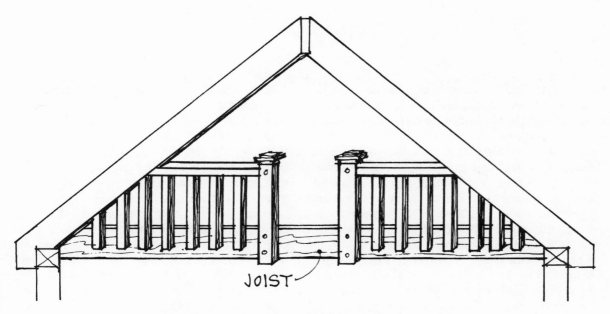

FIGURE 239. *One way to build a railing for a loft.*

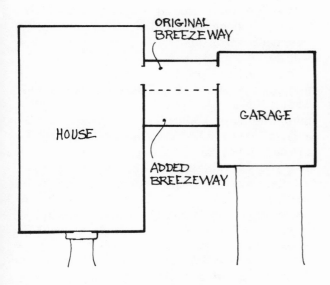

FIGURE 240. *Breezeway and garage make an L shape. If the breezeway is too narrow, it can be widened, but this entails virtual rebuilding of the structure.*

FIGURE 241. *Ways of heating an addition by the sun: south-facing sliders, French doors, and windows, in the gable or side wall. Skylights are good solar heating elements if the roof faces south.*

south (Figure 241). Some of the interior can be masonry to absorb and store solar heat, especially if the garage is located in a very sunny spot. Another type of heat to consider is a wood or coal stove, which wouldn't have to be very big to heat the space obtained in a garage conversion. Or you can extend your present heating system, or install electric baseboard heat. Electric heat is not generally recommended because it is expensive to operate, but with superinsulation your heating needs will be much less (see Chapter 3).

Interior finishes in your conversion are a matter of taste. Dry wall with a skimcoat of plaster is the most popular and serviceable, and paneling is a close second. Be careful with paneling; it can be too much of a good thing. One gable end of the garage, however, could be very effective if paneled, particularly if the paneling goes up to the peak of a cathedral ceiling. In a narrow breezeway paneling might be difficult because of the size of panel; wainscoting 2 to 2½ feet high along both long sides might be the better solution.

Converting a Basement

Conversion of a basement is mainly a matter of setting up frameworks to hold insulation and providing finish materials — walls, floor, and ceilings — and building partitions to divide the space as desired.

First, however, you must make sure the basement is dry. It must already be dry or must be made so, before you convert it. If it is damp from the condensation of water vapor on cool surfaces, this problem is easily solved by making the surfaces warm, with insulation; ventilating the area so that water vapor doesn't have a chance to build up; and dehumidifying when necessary and practical.

If there is moisture from seepage through walls or floor slab, or, worse yet, between walls and floor, you have a more difficult problem. Waterproofing a basement involves treatment of walls on the outside all the way down to the footings, installation of drainage pipes, and putting in sumps and sump pumps to take away water that threatens to enter.

There is one thing you can do, however, to reduce and perhaps prevent seepage through concrete foundation walls: paint the inside of the walls with a cement-based paint. This is a powder that is mixed with water. The walls are wetted down and the paint is scrubbed onto the concrete with a scrub brush. The paint is forced into the pores of the concrete to resist the encroachment of water.

Superinsulation of a basement is not as important as it is above ground; we have seen that the earth itself along the foundation keeps most of the basement warm. Standard insulation, therefore, will do.

Start with the floor. If the basement ceiling is at minimum height, such as 7 feet or less, you don't want to raise the floor very much, if at all. A minimum raising of the floor is obtained by applying a coat of hot tar, which must be done by a roofer. Or you can substitute roofing cement and do it yourself. Then put down 1 x 3 strapping on the flat side, on 12-inch centers. Between the strapping apply ¾-inch High R Sheathing or similar rigid insulation. Put 6-mil polyethylene over it and nail or glue ⅝-inch plywood to strapping and vapor barrier. Finish off the floor with oak strips or a plywood underlayment ½ inch thick and carpeting or resilient tiles on top (Figure 242). This much flooring will reduce the ceiling height by less than 2 inches.

If you can spare more than that, install 2 x 4s instead of strapping and embed them on their flat

Converting Garages, Breezeways, and Basements

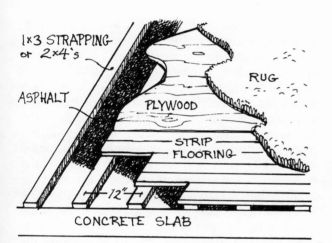

FIGURE 242. *A common and effective method for raising the floor of a basement. Air space between strapping is insulation. Further insulation can be installed by laying 3/4-inch Styrofoam or High R Sheathing between strapping.*

Walls are simple. Build a 2 x 4 stud wall and set it 2 inches from the foundation. Where the joists are at right angles to the foundation, nail the single top plate to them. Where they parallel the foundation, install 2 x 4 sleepers every 24 inches and nail the top plate to them (Figure 243).

At the places where the walls go above the foundation, install 6-inch unbacked insulation along the sill and against the header and stringer joists of the ceiling above (Figure 244). Then put 6-inch unbacked fiberglass into the wall so that it touches the foundation, and staple 6-mil polyethylene on the face of the studs as a vapor barrier. Wall finishes can be dry wall with a skimcoat,

sides into the tar or roofing cement. Put down 1½ inches of High R Sheathing in two ¾-inch layers. Then add your plywood and flooring. Carpeting in a basement or elsewhere below grade should be the type without rubber or foam backing, and the pad under it should be jute, not foam or rubber. A rubber or foam backing or pad will trap water vapor under it; if this condenses it could cause all kinds of water-related problems. The unbacked carpeting and jute pad will allow water vapor to migrate through and into the room, where it can be ventilated away.

If you don't use the hot tar or roofing cement on the floor, the strapping or 2 x 4s can be nailed or glued. Use concrete nails or construction adhesive. Remember that nails in concrete are often driven by an explosive nailing gun. This tool is best used by a professional.

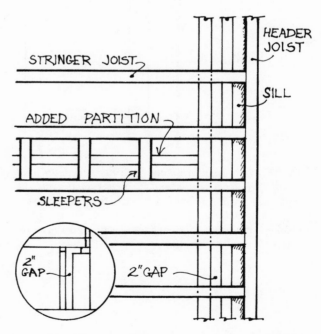

FIGURE 243. *Top plate of inner wall perpendicular to joists is nailed directly to the joists. Nail the top plate of a partition parallel to joists to sleepers nailed between joists.*

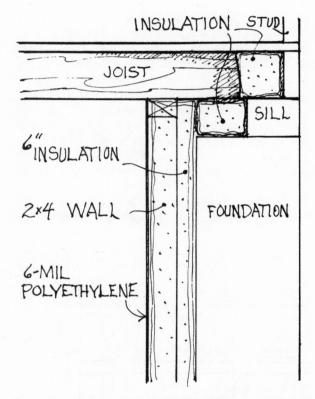

FIGURE 244. *Insulation goes against perimeter joist and sill. The inner wall stands 2 inches away from the foundation wall and is filled with 6-inch unbacked fiberglass, which touches the foundation. Six-mil polyethylene is attached to inner face of wall as an integral vapor barrier.*

taped and compounded plasterboard, or paneling. Paneling is glued to the plasterboard.

There is no full agreement as to whether you should insulate a basement ceiling when the basement is heated living space. We think it's a good idea, particularly if the heat in the basement is obtained with an auxiliary heating device such as electric radiators, or if the basement is not always going to be heated. With an insulated ceiling any heat will be retained. Also, if the basement is sometimes colder than the rest of the house, the insulated ceiling will keep the heat from the warm part of the house from escaping into the colder basement.

So put up 6-inch fiberglass insulation, the bottom of it lined up with the bottom of the joists and a vapor barrier, foil or paper, on the top, facing toward the ceiling.

A finish ceiling material is plasterboard, or a suspended type that uses a grid system and doesn't lower it at all. There are clips attached to the bottoms of the joists that permit a "zero clearance" ceiling.

A door that opens into the foundation will need an opening about 16 inches thick, counting the thickness of the foundation and the interior wall. A jamb is built of 1 x 10s or 2 x 10s ripped to the right width, and a piece of trim or molding to cover the joint. Build the door frame on the floor, insert it in the opening, and nail it to the interior walls (Figure 245). Depending on where the joint between the jamb members is, you can use it as a stop for the door.

Windows are treated in the same way. Build a frame of 1 x 10s on all four sides, making an open-ended box, and insert it in the window opening; nail it to the wall ends (Figure 246). If a window is already in place, your frame is of a different depth, but it is still inserted in the opening in front of the

Converting Garages, Breezeways, and Basements

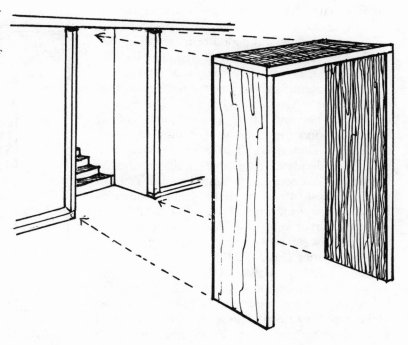

FIGURE 245. *Exploded view of door frame devised to fit into opening in foundation wall. The door fits into this frame, and the casing fits onto the facing edge of the frame, flush against the inner wall.*

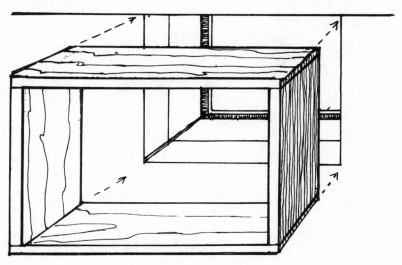

FIGURE 246. *Exploded view of window frame that fits into window opening. After the frame is set into the opening, the casing is nailed around its perimeter, flush against the wall.*

window. There will be enough clearance for the window to open.

Outside window and door storms are available, and inside storms or thermal shutters can be built and installed, to complete the superinsulation.

Some incidentals about basements: the furnace or boiler can be sealed off with walls. This cubicle should be big enough to permit plenty of air volume around the units so that they can burn fuel safely. They need air to burn; 10 by 10 feet should be adequate. Insulate the walls around the furnace or boiler cubicle to control sound. You can provide outside air to the fuel-burning appliances by running a 4-inch metal duct or a flexible one such as a dryer vent hose, from a window or an opening between two joists to the area in front of the fuel burner (Figure 247). The duct can also end at the ceiling. This way the fuel burner gets its own outside air and will not suck air from the living areas of the basement or the main house. Make sure the outside end of the duct is covered with screening and ¼- to ½-inch hardware cloth (steel mesh) to keep out insects and vermin.

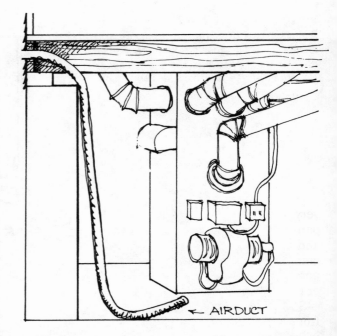

FIGURE 247. A duct from outside is aimed toward furnace or boiler (any fuel-burning appliance) to provide air for combustion.

chapter 20

Living on the Outside

Porches, Decks, and Patios

Despite the lament that modern houses are not very interesting because they have no porches or pantries, porches are passé. Nowadays houses and additions generally have only stoops—small landings with steps, roofed or not roofed. The full porch of yesterday has been replaced by the deck or patio of today, sometimes roofed and screened.

Still, there must be some kind of area at the front and back entrances, and an ell-type addition will need at least one of them. If it serves as a front entrance, it can be a stoop or small porch; if it is only a back door, it can be just a stoop.

The best kind of stoop-porch, at least for durability and low maintenance, is a concrete slab with concrete steps, made more attractive if faced with bricks. A ground-level entry is good to have if there is any part of the house or addition with the floor at ground level. The slab for this entry, by the way, should be at least 6 inches above grade.

A concrete stoop slab needs footings and a foundation. Sometimes this is built as part of the foundation of the addition itself. If it is not, it should have its own footings and foundation (see Chapter 4), separate from the foundation of the addition (Figure 248). The stoop foundation should not be tied into the addition; if it should move, you will have all kinds of problems. The concrete stoop, set on proper footings and foundation, probably won't move; and since the house or addition also has, or should have, its own foundation, there should be little movement of either and equally little danger of a large gap opening up between stoop and house—a small one, maybe, which can be filled with mortar, but not a big one.

The slab is poured on top of 6 inches of gravel inside the foundation. Never pour concrete directly on filled earth; even with 6 inches of gravel the earth should be undisturbed and tamped very thoroughly. Steel reinforcing mesh or rods will prevent any vertical movement of the 4-inch slab if the earth underneath drops even slightly (Figure 249).

A concrete step or steps should be built as part of the slab and have their own footings and foundation. Build forms for steps and slab (Figure 250), following the step formula: two risers and one tread each 25 inches. The steps can be solid concrete on top, but their interiors can be rubble or crushed stone. They should pitch ¼ to ⅜ inch away from the house, and all the concrete should be finished with a wood float for a rough, nonslip finish.

It is best to set the stoop slab one step down from the threshold of the addition; anything less would be awkward and cause stumbling. Also, if the threshold were very close to the slab, water could back up under it and eventually cause decay of the wood members. To accommodate an out-

Living on the Outside

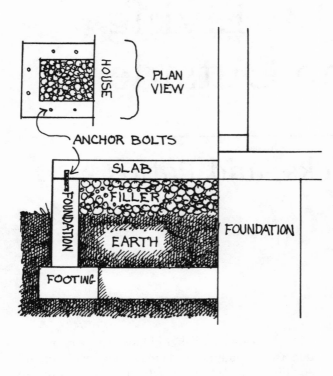

FIGURE 248. *A concrete slab for a stoop-porch needs its own footing and foundation. It can butt up against the addition but should not be tied into it unless its foundation is part of the addition's foundation.*

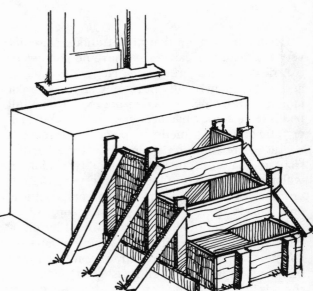

FIGURE 250. *Forms for concrete steps (and slab, shown finished) must be strongly reinforced because of the great pressure of fresh concrete.*

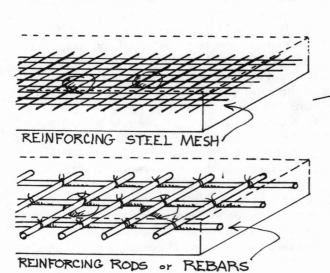

FIGURE 249. *Reinforcing mesh or rods in a slab of concrete. Stones are placed under them so they will remain in the middle of the slab's thickness as the concrete is poured. Bars are tied together with steel wire.*

Porches, Decks, and Patios

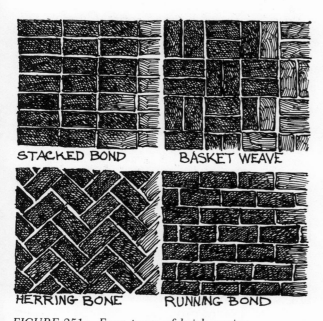

FIGURE 251. *Four types of brick paving.*

swinging screen or storm door, the stoop slab should be at least 4 by 4 feet. It should be bigger if there will be a roof over it, in order to be in the proper proportion to the addition.

You can buy a ready-made concrete step-stoop combination, which needs only to be set on its own foundation.

Brick for the stoop and steps will improve the looks of the entrance greatly. To accommodate the thickness of a brick (2½ inches if it is set on its wide side, 3¼ inches if set on its narrow side), you should lower the slab appropriately. Allow ⅜ inch for the mortar joint. The bricks can be laid in fresh concrete by embedding them slightly. They can be mortared at that point, or put in with gaps and the mortar applied after the concrete sets. If you put bricks along the side of the concrete stoop or porch, you should pour the concrete foundation so that there is a shelf at least 4 inches wide along its periphery. This permits the side bricks to sit on the foundation at the earth line.

The bricks on the stoop can be laid in any of several patterns (Figure 251). Edge bricks should be placed to form a border, with the ends of the bricks overlapping the slab by ½ to 1 inch, as a drip edge. At the corners a patio block or decorative stone should be cut to fill the square, preventing the nearby bricks from falling out (Figure 252). It's a good idea to set the edge bricks on their narrow sides.

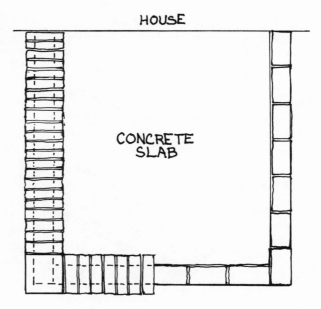

FIGURE 252. *Edge bricks on the perimeter of the concrete slab are set on their narrow edges and overlap the concrete by ½ to 1 inch. Corners are stone or large pieces of concrete. If bricks were set in a corner, their wide edges would show, and they tend to loosen when they are close to an edge.*

Living on the Outside

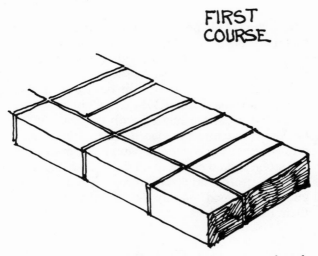

FIGURE 253 A. *First course of a brick step, with each brick set on its wide side. This configuration allows a 10 to 10½-inch tread.*

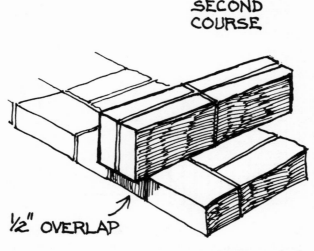

FIGURE 253 B. *Second course of a brick step, with each brick set on its narrow edge, overlapping the lower course by ½ to ¾ inch. This configuration allows a riser of 7 inches.*

Steps can be made almost entirely of brick. The first course is a row of bricks set in a ½-inch mortar joint on their wide sides so that they are parallel with the length of the step. They are placed to form a tread of 10½ inches (Figure 253 A). The tread itself is made of a second layer of bricks set on their narrow sides, perpendicular to the length of the step (Figure 253 B) so that their ends extend over the row below by ¾ inch. This forms a drip edge. The space behind the two courses is filled with concrete.

To make the second step, repeat the procedure. The base bricks are laid 10½ inches in from the edge of the first step. Continue upward to the height of the top stoop, filling in behind the brick steps with concrete or brick (Figure 254).

When you work with brick, keep your mortar wet enough so that you can butter the edges of the bricks as you go along, but not so wet that it spreads too thinly or is too runny to support the bricks on top of it. If you have embedded the bricks in concrete and left the joints open, the mortar that you use to fill those joints should be almost dry, just crumbly. Then, when you pack the mortar into the joints, it won't get all over the face of the brick. Pack it thoroughly and completely; you will be amazed at how much mortar you can stuff into a joint to make it flush with the surface face. After the mortar has set for ten or fifteen minutes, strike the joints by dragging a pointing tool—an S-shaped steel rod—over the mortar to create a slightly concave joint. The more you work mortar the wetter it gets, so avoid overworking.

Mortar that does get on the face of the brick can be removed by treating it with muriatic acid. Dilute 1 part acid with 2 parts water and pour it generously over the face of the brick. It will fizz up on the mortar; when the fizzing stops, scrub with a broom and wash off with lots of water from your garden hose. Repeat if necessary.

Porches, Decks, and Patios

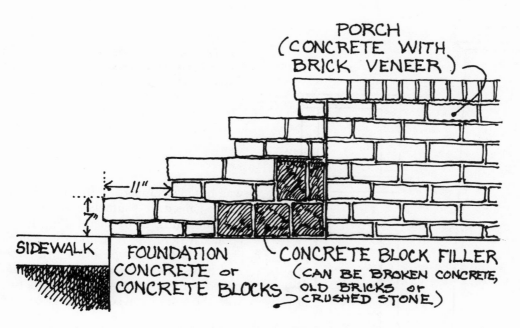

FIGURE 254. *Cutaway profile of brick steps shows the position of the bricks forming the steps and the material backing them up. Each end of the steps will have a veneer of brick to match the side of the porch.*

FIGURE 255. *A porch with a gabled roof.*

New or used bricks are good for porches, but be sure that they are water-struck, or at least hard, bricks, designed for such usage. Common brick, new or (especially) used, will not stand up to weather when laid horizontally.

We should point out that brick work is tricky, and unless you have total confidence in your ability, you might prefer to hire a mason and act as hod carrier and general factotum, not master bricklayer.

You might like to put a roof over your stoop or porch (Figure 255). Since a front porch is not usually walled in, the roof is held up by columns or pillars. You can buy round or square columns of various sizes, made of aluminum or wood, complete with bases and capitals. But you can also make your own.

Nail nominal 1-inch boards into a long, open-ended box. It can be 4 by 4 inches, 6 by 6, or larger, depending upon the size of your porch and the length you want for the columns. Make sure the box is square, and taper it slightly for a graceful look (Figure 256). Capitals and bases can

Living on the Outside

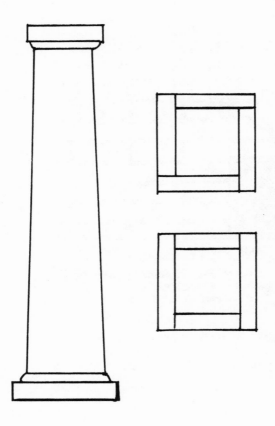

FIGURE 256. *A square pillar or column made with ¾-inch boards, each board tapered ½ inch for a graceful look. End views show two ways of nailing the boards.*

wall but also supporting the roof. The roof can be flat or gabled. It is attached to the addition when the sheathing is in place. Flashing should be applied where the roof and the wall meet.

Porches can also be fitted with wood railings and balusters. A railing consists of two 2 x 4 or milled rails spanning the space between posts, with the bottom rail set on blocks to keep it off the

be made by cutting nominal 2-inch lumber 1½ inches larger than the column on all four sides, toenailing through the column sides, and finishing with a ¾-inch round between the column and the base or capital (Figure 257). All joints should be caulked. Attach the base to the porch by drilling through it and the corner stone (or brick or concrete) and inserting a steel rod. This will secure the column laterally; there is no need to attach it directly to the porch floor. The weight of the roof is enough to hold it.

Where the pillars touch concrete, or even wood for that matter, they should be treated with wood preservative and raised slightly to allow drainage. If you have a solid 4 by 4 or 6 by 6 post, you can use a post bottom or post base or similar fastener. Or, raise the post slightly by driving bolts into the bottom and using the bolts as feet. To keep the post from moving, anchor a bolt or put an anchor bolt in the concrete, or drill a hole in the stone or brick and insert a bolt, any of which will stick up into a hole in the bottom of the post base (Figure 258).

To complete the roof, the inside support can be pilasters, which are half-pillars set up against the

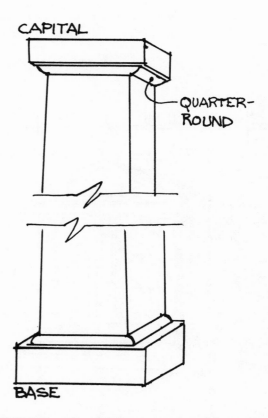

FIGURE 257. *The simplest capital and base for a square pillar is a 1½-inch board at top and bottom, with ¾- to 1-inch quarter-round as a finishing touch.*

concrete. The rails are separated by balusters—vertical spindles—made of 2 x 2s or milled ones 1¼ by 1¼ inches, or fancy turned ones. The height of the balustrade (the whole unit) is 24 to 30 inches, depending on the size of the porch.

If the roof is flat, a low balustrade can be installed along its edge (Figure 259). It should be lower (18 inches), in keeping with the scale of the porch. To relieve the possible monotony of a parade of balusters all in a row, you can interrupt them with an X in the middle of the run, made of the baluster material. The X can be varied by superimposing a cross on it, making a design known as a star. Of course, fancy turned balusters will not work if you want to make the X.

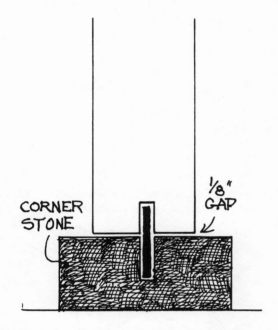

FIGURE 258. *A solid piece of wood, such as a 4 × 4 or 6 × 6, can be a column or pillar, and is secured with a pin to keep it from moving laterally. Depth of hole and length of pin determine gap for drainage.*

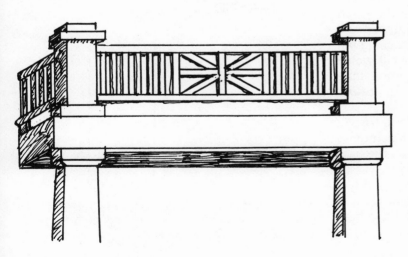

FIGURE 259. *Flat roofed porch with a small decorative balustrade.*

Living on the Outside

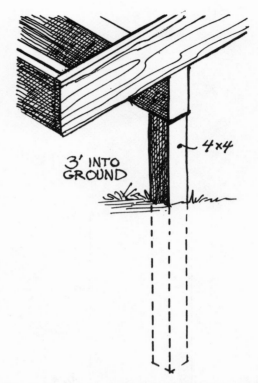

FIGURE 260. *A small porch or wood deck can be built on wood posts set 3 feet into the ground.*

WOOD PORCHES

A wood porch is simple to make; the construction is similar to that for a wood deck, the modern replacement. There is little difference between the two, except perhaps for size. Use pressure-treated wood for both.

A small porch can be built on wood posts (4 x 4 or 4 x 6) set 3 to 4 feet into the ground. As a matter of fact, a larger deck also can be built this way (Figure 260). Or you can use concrete Sonotube piers, set 3 feet into the ground. No footing is required (Figure 261).

A doubled 2 x 10 beam is set on top of the posts (Figure 262), using special fasteners, or a 2 x 10 or 2 x 12 is set on each side of the posts, sitting on 2 x 4 vertical cleats nailed to the posts (Figure 263). In the case of concrete piers, the doubled beam (or one a nominal 4 inches wide) is set on the pier with the help of a fastener (Figure 264).

Joists 12 feet long are set from a ledger on the addition or house to the beam, or beyond it (Figure 265). Thus the deck can be up to 12 feet deep and as wide as you want to make it. Flooring is made of 2 x 4s set ⅜ inch apart for drainage (Figure 266). If you plan to screen in the porch and use a roof, the floor should be tongued and grooved fir flooring, usually 1 x 4s. You could put a screen under the spaced 2 x 4s, but it would collect debris and be impractical. A low railing around the porch or deck is attractive; Figure 267 shows several ways to set one up.

Putting a roof over a deck requires the base of the deck to be as strong as a regular structure, so the base must be built the same way as the base for the addition, with piers, beams, and joists (see Chapters 4 and 5). Studs can be set on 3- or 4-foot centers and screens put between them (Figure 268). If you put in cross-members, put them high or low enough not to block your view.

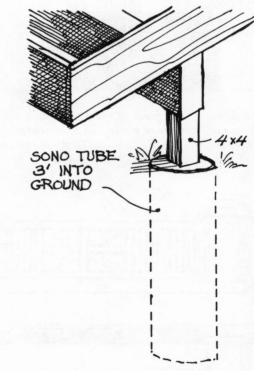

FIGURE 261. *A porch or deck can also be set on concrete piers sunk 3 feet into the ground, with a wood post sitting on top of the pier.*

Porches, Decks, and Patios

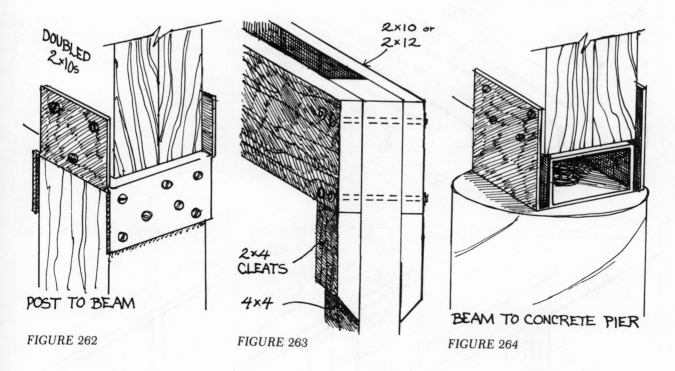

FIGURE 262. How a fastener secures a doubled beam to a post.

FIGURE 263. Another method of securing a doubled beam on posts. Here one section of the beam is bolted on one side of the post, the second section to the other side. The beam is also supported by 2 x 4 cleats nailed to the post.

FIGURE 264. How a beam is secured to a concrete pier with a steel fastener.

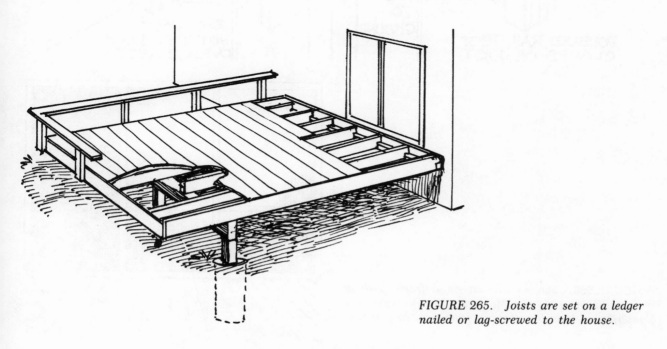

FIGURE 265. Joists are set on a ledger nailed or lag-screwed to the house.

Living on the Outside

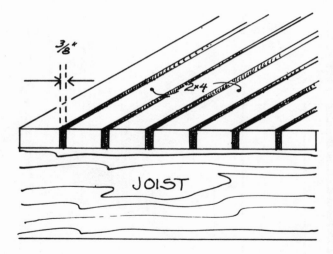

FIGURE 266. Deck boards (here 2 x 4s, but the technique works with 1 x 4s, 2 x 6s, and 1 x 6s, etc.) are spaced ⅜ inch apart for drainage.

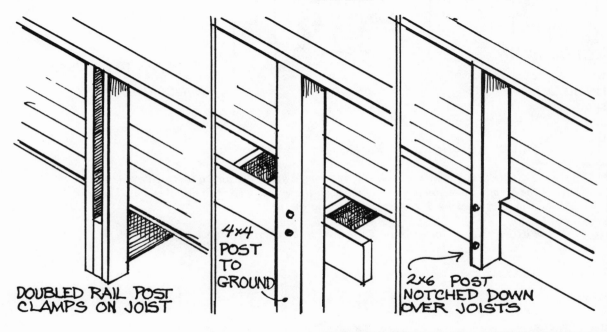

FIGURE 267. Various treatments for railings on a wood deck.

FIGURE 268. Screens are set on a frame with a maximum span of 4 feet. Make sure cross-pieces do not interfere with the line of vision from a seated position.

Porches, Decks, and Patios

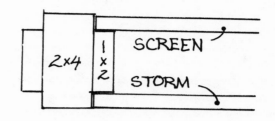

The easiest way to mount screens is to nail 1 x 2s along the sides of the 2 x 4 studs. This will form an inner lip for the 1 x 2 or aluminum frames of the screens to rest on (Figure 269), as well as an outer lip to accommodate wood storm windows or 1 x 2 frames with acrylic or polyethylene stapled to them. Set the 1 x 2s so that the storms will be flush with the studs and can be secured with butterfly fasteners (Figure 270). This way you will keep snow and rain off the porch or deck in the winter, and the closed-in deck will act as a buffer against the cold.

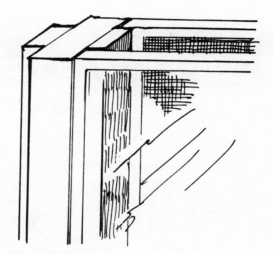

FIGURE 269. *One way to create a lip with 2 x 4s and 1 x 2s for fitting a screen on the inside of the frame and a storm (temporary, such as polyethylene on a wood frame; or permanent, aluminum or wood frame with glass or acrylic glazing) on the outside.*

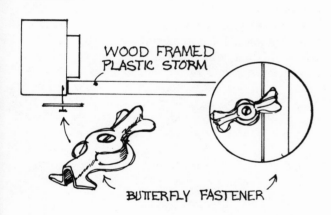

FIGURE 270. *Old-fashioned butterfly fastener holds wood-framed plastic storm window.*

PATIOS

With your addition done, you might well want to put a patio in the inside corner of the *L* shape formed by house and addition. Or it might be a good time to put a patio anywhere.

A patio is generally set on the ground and is made of masonry instead of wood. It acts as a flat surface for the placement of furniture, reducing or preventing wear and tear on a grass lawn, or serving as a pavement under a tree where grass won't grow anyway. The best patio material is bricks laid in sand (Figure 271). This is not inexpensive, but it is sturdy, easily maintained, stylish, and adaptable to various locations, designs, levels, and specifications (as sidewalks, for example). Also, bricks in sand—in fact, any materials on top of or in sand—are good under trees and in other areas where there are growing things because the sand allows rainwater to filter into the ground.

First, determine the size and shape of your patio and set up the borders, which can be landscape timbers (use pressure-treated lumber), concrete or granite curbing, patio blocks (8 by 16 or 12 by 24 inches), or soldier bricks, set into the ground

Living on the Outside

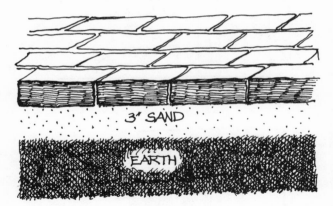

FIGURE 271. Brick in 3 inches of sand makes a permanent, good-looking patio.

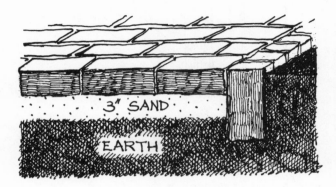

FIGURE 272. Soldier bricks set the long way create a border for a patio that keeps other bricks from spreading.

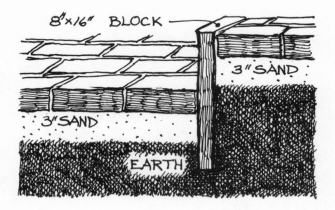

FIGURE 273. Where there is a step in the patio, or where the patio is higher than the surrounding lawn, a concrete patio block set with its 16-inch length partway into the earth works as an edger.

the long way (Figure 272). Dig into the ground so that the tops of the border pieces are level with the surface of the patio. Any border material will work nicely if the patio is on the same level as the surrounding lawn, but if it is above the level of its surroundings, landscape timbers or patio blocks are the best edgers. Sink the 16-inch patio blocks the long way into the ground, leaving enough of each above ground to match the height of the step or differential between levels. If you use the 12 by 24-inch blocks, you can sink them the 12-inch length if the step is no higher than 3 or 4 inches (Figure 273).

In the interior of the patio, dig down 5 inches. This will allow for 3 inches of sand (you can also use stone dust, which packs down better) and the 2¼-inch thickness of the brick. Those numbers total 5¼ inches; either depth is fine. You can use a 2 x 6 or 1 x 6 as a guide in digging. Such a board is actually 5½ inches wide, so when its top is lined up with the top of the border, you have excavated 5½ inches. That's a little deep, so add some sand.

Spread the sand evenly and tamp it down with a roller or tamper (use your shod feet or a homemade tamper, a 12-inch-square piece of ¾-inch plywood nailed to a 2 x 3 or pipe handle). Wet down the sand and retamp; let it sit for a few days and tamp again, adding or removing sand as necessary. The more thoroughly you tamp and prepare the sand and let it settle with time, the easier it is to lay the bricks.

The bricks can be laid in any pattern; the common, or running, bond is easiest and best (see Figure 250). A basketweave pattern is handsome, but not practical with most bricks because it requires wide joints to make the bricks line up. A herringbone is also nice, but needs frequent cutting at the borders. Also attractive are concentric squares or rectangles (Figure 274).

FIGURE 274. Concentric squares make a good brick pattern in a patio, with a minimum of cutting.

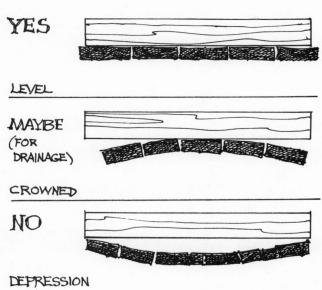

FIGURE 275. A straight board will help determine whether a row of bricks is straight and level.

Be sure to use hard bricks, such as paver bricks, or water-struck bricks. Common bricks, particularly used ones, will not stand up to the weather on a horizontal surface.

Butt the bricks as closely together as possible. You will use 5 bricks to the square foot. Never mortar bricks set in sand. When they are all in position, sweep fine sand or stone dust into the joints. Wet down and repeat. Eventually you will just about fill the joints and the sand or dust will tend to harden. You may have to do this once a year or so. Depending upon how well you tamped the sand, you will probably have to pick up some of the bricks in five or so years, add a little sand, and relay them in the right position.

As you lay the bricks, check them with a level frequently. Check every row because it is very easy to get out of level. You can stretch a string from border to border, but this isn't wholly reliable because you can inadvertently raise the string as you work. So a long board spanning the space from border to border is a better check on levelness (Figure 275). Note that the patio can have a slight crown in the middle for drainage, but levelness is preferred because it will drain anyway. Be sure that individual bricks don't hump up or ride low—your board will tell you. If the border-to-border distance is too far for one board, set up a small pier at the midpoint for one end of the board to rest on. Levelness must go in two directions: the length of the bricks and at right angles to the length.

If you build the patio or sidewalk in the sun, lay perforated black plastic or roofing felt under the sand. This will prevent weeds from growing up between the bricks. In the shade, this is a minor problem and weeds can be readily pulled out by hand. The perforations in the plastic permit rainwater to drain into the earth rather than just sit in the sand.

part III

Don't Give Up The House

Superinsulating an Existing House

Chapter III

Don't
Give Up The House

Superimposing on Existing House

chapter 21

The Easy Parts

Attic Floor and Basement Ceiling

Why bother with superinsulating an existing house? Why not just tighten it up reasonably, maybe add some solar heat, and hope for the best?

That is one way to proceed. Superinsulation is another. It's not just a matter of adding tons and tons of insulation. That will help, of course, but just as important is making the house airtight. To do this you need a complete vapor barrier. So wherever you add insulation, you will also put a vapor barrier toward the heated side of the house.

Superinsulation and full vapor barriers put in an existing house will not be as complete and effective as those installed in a new house as it is being constructed, but they will go a long way toward freeing you from high heating costs. Think of this: oil or heating bills of perhaps $100 a season, because once the temperature in a superinsulated house reaches 70 degrees, it can stay that way through intrinsic heat (from people's bodies, cooking, and so on) with very little additional heat needed.

If you had the work of superinsulating your house done by hired help, it would probably be too expensive and the payback time too long. For instance, if you paid $10,000 for the job and saved $1,000 a year in fuel, the payback time would be ten years. On the other hand, if you do the work yourself, the cost is cut in half and the payback time becomes five years or less, which is certainly worthwhile. Moreover, superinsulating is something that you can do yourself.

Start where it is easiest: the attic floor (first-floor ceiling in a one-story house). If there is no insulation there at all, your job is even easier. You can put a vapor barrier on the inside surface of the ceiling between the joists, but this has occasionally caused problems. Moisture sometimes gets trapped between the vapor barrier and the ceiling and condenses; if that moisture sits there long enough, it can cause decay of both wood and plaster. So it is better to paint the ceiling with a vapor-barrier paint. It is white and both a sealer and primer and is a flat paint. It can pass easily as a ceiling paint, or eventually ceiling paint can be put on over it.

Then lay 12 inches to 18 inches of unbacked fiberglass roll or batt insulation in the attic floor. The first 6 inches or so goes between the joists; try to get as close to the tops of the joists as possible. Then another layer goes on top of the first, at right angles to the joists. Thus you will cover both joists and first layer (Figure 276). You may have to combine different thicknesses to get the first layer even with the tops of the joists, or at least within ½ to 1 inch of the tops, higher or lower. The reason for trying to match up with the joist depth is that any greater gap will make the second layer of insulation undulate as it rises or

The Easy Parts

falls over the joists, and this can leave spaces, which reduce the effectiveness of the insulation. The insulation must be tight against each unit and must be complete.

Twelve inches of fiberglass will give an R factor of 38 to 40, pretty good. Another 3½ inches, bringing the thickness up to 15½ inches, will raise the R factor to 49; quite good. Anything more is spectacular.

When you put insulation in the attic, do not put it in the eaves; as we've seen, the air passages from the eaves up between the rafters should not be blocked. If necessary, install a barrier of wood boards between the rafters or nailed to them, to hold back the insulation from any air passage. If the eaves are not kept free of insulation, serious moisture problems can occur, and they are more likely to occur than not.

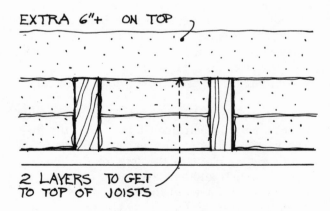

FIGURE 276. *Insulation can be adjusted so that it comes up to the tops of the joists, when set between joists in an attic floor. Then more insulation is added, at right angles to the joists. This prevents gaps in the insulation and covers the joists, which are a potential source of heat loss.*

FIGURE 277. *In cases where a ceiling has a coat of vapor-barrier paint, it must be scored to render it inoperable before a new insulated ceiling is installed under it.*

If there is some insulation already in the attic floor, add enough unbacked fiberglass to bring the total thickness up to 12 to 18 inches. Lay at least one layer at right angles to the joists.

You may want to put up a new ceiling below the attic floor, but you do not want more than one vapor barrier on any one surface. Moisture could be trapped between the two vapor barriers. So put 6-mil polyethylene on the old ceiling and then put the new plasterboard or plaster ceiling directly over it. With a dropped ceiling, you can install a vapor barrier on the inside of it and put as much insulation on it as will fit, but if there is a vapor barrier on the attic floor, it must be removed. If that is not possible, do not add insulation or a vapor barrier above the dropped ceiling (that is, between old and new ceilings). With a saw, you can score any vapor-barrier paint that is on the old ceiling to make sure water vapor passes through it (Figure 277).

Basement Ceilings and Walls

Basement ceilings are quite easy to do if they are not plastered. A lot of them are, to delay any fire in the basement from breaking through the floor. This danger is less likely today, with modern automatic heating equipment, but if you use a coal or wood-burning appliance, the hazard is there.

Dead-air space in an intact plaster ceiling is a good insulator. To find out how good, determine the temperature in the basement. If it is under 50 or near 40 in the dead of winter, the insulating factor of the dead air is probably pretty good, and you are not getting too much loss from the heated part of the house. But if it is warmer than, say, 50 or 55 in the basement, chances are the insulation is not too good. The heat may also come from the furnace or boiler or from the heating air ducts or hot-water pipes. In that case, make sure your furnace or boiler is insulated (consult a professional) and insulate the heating ducts and pipes (see Chapter 14). With an intact plaster ceiling you can have insulation blown in.

Broken plaster should be patched; or you can take it down and insulate with 6 inches of fiberglass insulation. Put a vapor barrier on top, toward the heated part of the house (Figure 278). Any air space between vapor barrier and floor is a bonus. If you have no plaster ceiling, insulate between the joists as described in Chapter 19.

Basement walls can be insulated according to the technique in Chapter 19. And, as we pointed out in that chapter, if the basement is not always used for living space or is separately heated, you should certainly insulate the ceiling.

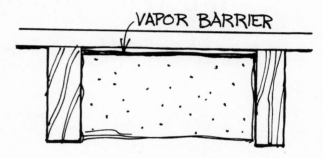

FIGURE 278. *In a basement or crawl-space ceiling, 6-inch fiberglass insulation is installed between joists, with a vapor barrier toward the heated part of the house (the floor).*

chapter 22

The Hard Part

Superinsulating Existing Walls (and Windows and Doors, Too)

WALLS

Superinsulating existing walls means achieving an R factor of 30 or greater. There are two ways to do this: by applying a thin but unusually high R material to the wall itself, or by adding an extra wall on the inside. And there are several different methods to achieve these ways. None of them is easy. Walls are the least convenient part of the house to superinsulate because you are necessarily robbing the existing rooms of a small amount of space, and willy-nilly many things have to be removed. But it is worth it.

First things first. Let's say you decide to put the rigid high R material on the existing walls, without removing their surfaces. If there is no insulation at all in the walls, some must be put in. You can have fiberglass blown into the wall cavities. That, plus sheathing, siding, and interior wall finish, will give an R factor of 9 to 11 or so. Or you may already have insulation in the walls that amounts to about the same R factor.

Now you must find out whether there is a vapor barrier just inside the inside wall surface. If there is, it should be eliminated. If there isn't, determine if the wall has been painted with a vapor-barrier paint, or is papered with vinyl-coated, canvas-type, foil, or Mylar wallpaper. Any of these will have to be removed, because added insulation must go on before any vapor barrier, and the vapor barrier must go only over or under the final interior wall finish (Figure 279).

You should also remove any window and door casings, baseboards and shoe moldings, and other interior trim. The reason is that the added insulation must go right up to the window and door jambs and cover the area behind the casings, and go down to the floor itself. Save the material you remove, however; it can be reapplied. And there is an advantage here; you can caulk all the casings when you reinstall them (see below).

That is one way to prepare for putting on high R insulation. Next let's consider actually taking off wall surfaces first. This procedure involves considerably more work, but it makes possible the installation of fiberglass batts between the studs.

Taking off existing plasterboard is done by brute force. You drive a pinch bar into the wall and pull. The plasterboard will come off in great, heavy chunks that should be dumped out of a window. You can use the same method to remove walls plastered on Rocklath (small panels of plasterboard), but it takes a good deal more force. When your wall is old plaster mounted on wood lath, do it this way: cut long strips of the plaster, about 12 inches wide, with a chisel and hammer, utility knife, or other sharp instrument (it will dull quickly), and pull off each strip (Figure 280).

Superinsulating Existing Walls (and Windows and Doors, Too)

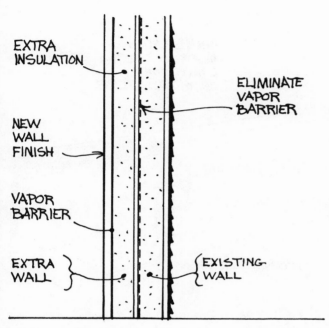

FIGURE 279. *An added wall will double insulation value. It is essential to eliminate any old vapor barrier on the inside of the existing wall, or to pierce it so that water vapor can pass through it.*

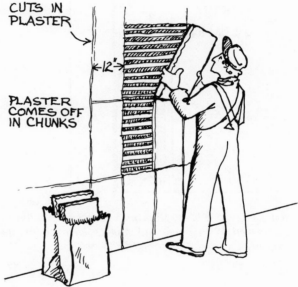

FIGURE 280. *How to remove old plaster and dispose of it. This eliminates handling the heavy plaster several times. The pieces fit nicely into a grocery bag.*

Break the strips up into 1½- to 2-foot lengths, bag them, and take them to the dump. If you can toss them directly into a dumpster or pickup truck backed up to a window, so much the better. The point is that plaster is very heavy, so you want to handle it as little as possible. Then remove the wood lath.

With the wall open, you can now install the unbacked fiberglass between the studs, if there is little or no insulation there already. It will stay in place by friction fit (Figure 281).

What about heating units along the walls, using either of the procedures just described? Any radiator should be removed when you take off the plaster. Or you can put insulation behind it and install a surface of sheet aluminum behind it (Figure 282). When the new insulation is put on the wall, it in effect surrounds the radiator and

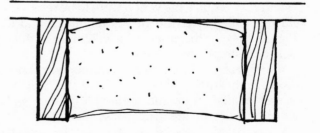

FIGURE 281. *Unbacked fiberglass insulation remains between studs by friction. Inside wall finish will also hold it in place.*

195

The Hard Part

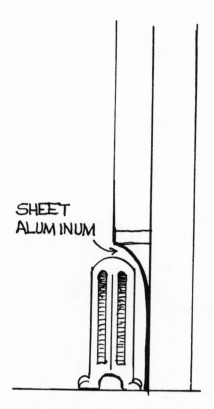

FIGURE 282. *Sheet aluminum behind a radiator reflects heat. With a new wall in place, surrounding the radiator, the radiator is in effect recessed.*

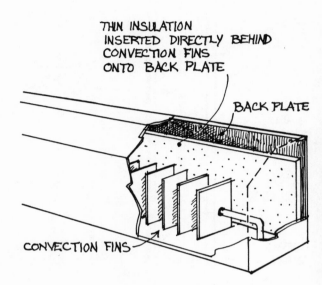

FIGURE 283. *Thin insulation is applied to the back plate of a steel baseboard unit. Combined with the reflective surface, it is effective.*

recesses it. This is okay because there will be a fair amount of insulation behind it.

A baseboard heating element can be recessed, too, although the baseboard and the wall material behind it are left intact. When the baseboard units are steel, a special adhesive-backed insulating reflector can be applied to the back plate of the unit (Figure 283), in addition to new insulation behind the baseboard. A hot-air duct and vent system may require the vent fixture to be removed and reinstalled after the wall is rebuilt around it (Figure 284); or the vent and sometimes the duct itself can simply be moved.

At any rate, recessed heating elements will not lose their heating ability, and there will still be enough insulation behind them so that they will not be sources of great heat loss.

A baseboard radiation heater can be moved instead of recessed. This is relatively easy; it is simply a matter of making the T connection at each end of the unit horizontal instead of vertical,

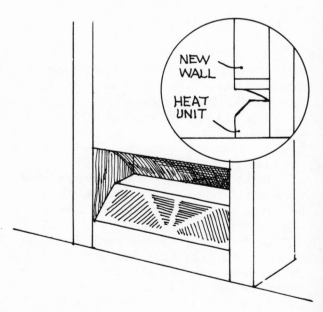

FIGURE 284. *Baseboard hot-air vent is recessed by building the new wall around it.*

Superinsulating Existing Walls (and Windows and Doors, Too)

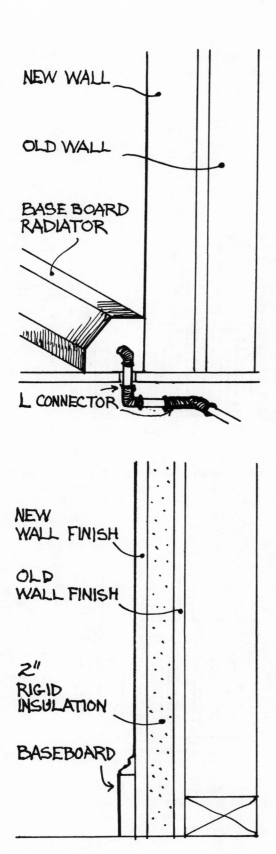

and attaching an L connection to move the pipe forward and permit the unit to come forward, too (Figure 285). Radiators can be treated in the same way. If there is no access to the floor above, all the turns can be made above the floor. Old radiators with cast-iron pipes or pipes other than copper may require the services of a plumber to move them. But these adjustments are certainly feasible, and there have to be some compromises made when superinsulating an existing house.

Now for the extra insulation on the inside wall. Two-inch High R Sheathing, Thermax, or other brand of polyisocyanurate has an R factor of more than 15. Added to the 11 or so in the existing wall, it will up the R factor to 26, quite reasonable, and will bring in the wall surface only about 2½ inches (Figure 286). If you haven't removed the old wall, you can glue the insulation with construction adhesive directly to the wall surface. If you have removed the old surface, nail the insulation to the studs. This material has an aluminum-foil face and back, which is fine, since the entire unit is its own vapor barrier. Seal joints with duct tape.

The material is, however, flammable, so it should be covered by adequate plasterboard as a finish; ⅝ inch is recommended and is sometimes required by code. This applies whether the existing wall is removed or not. The plasterboard can be nailed right through the insulation into the studs, using extra-long dry-wall nails.

FIGURE 285. L connector diverts hot-water heating pipe to the new location of the baseboard radiator, in front of the new wall.

FIGURE 286. The minimum amount of space is lost by adding 2-inch rigid insulation and a new wall finish. Woodwork is also relocated onto the new wall. Old wall finish is left intact.

The Hard Part

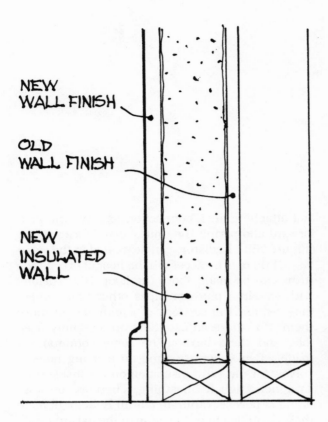

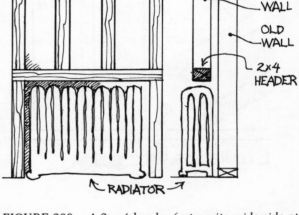

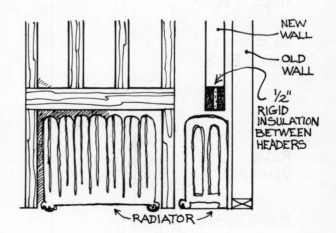

FIGURE 287. *A double wall in an existing house: the new wall is filled with insulation and a new wall finish is put on. Old wall finish is left intact.*

FIGURE 288. *A 2 x 4 header (set on its wide side at the bottom of studs, and supported by studs on each side of the radiator) supports a new wall over a radiator.*

FIGURE 289. *If a radiator is particularly long, the header is like that over a wide window: two 2 x 6s or 2 x 4s, on their narrow edges, with insulation between them.*

The second way to add insulation to walls is to build a separate wall of 2 x 4 studs and finish it off with plasterboard or some other dry-wall system (Figure 287). You can put 3½ inches of fiberglass in that wall, adding another R factor of 11, and making the entire wall's R factor 22. But that is not enough. Therefore, if you have the space, set the new inner wall 2 inches away from the existing wall and install 6 inches of fiberglass to bring the total R factor to 30. Before putting on the wall finish, cover the insulation with 6-mil polyethylene as a vapor barrier, sealing the entire wall.

If you have to build the inner wall around a hot-air vent or radiator, put a 2 x 4 header across the span above the radiator to hold the studs above it and make the span sturdy (Figure 288). For a particularly wide radiator, you may have to use two 2 x 4 or 2 x 6 headers, separated by ½-inch High R Sheathing (Figure 289).

Treatment of a baseboard heating unit running the entire length of a wall is another matter because in effect you will build the wall above it. So nail a 2 x 4 to the existing studs as a cleat, to hold the 2 x 4s of the new wall. These 2 x 4 studs will have to be notched. After the wall is built, nail

Superinsulating Existing Walls (and Windows and Doors, Too)

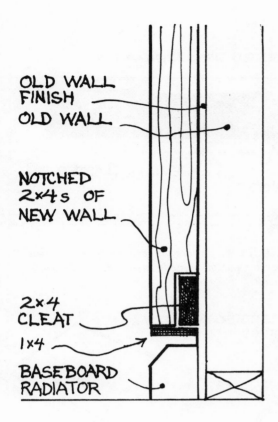

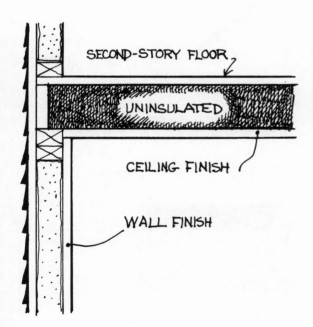

FIGURE 291. *The ceiling space on the first story of a two-story house is uninsulated. This is okay except where the ceiling ends with a perimeter joist, which is the only thing (plus sheathing and siding) separating this space from the cold outdoors.*

FIGURE 290. *For a full-length opening over a baseboard radiator, the studs are notched to fit over a 2 x 4 cleat nailed to the studs of the existing wall.*

a 1 x 4 bottom plate along the bottoms of the studs (Figure 290).

A seriously neglected area in the house as far as adding insulation is concerned is where floor joists of the second story sit on top of the first-story walls (Figure 291). If this area is not insulated, it is virtually defenseless against the infiltration of cold; only the header or stringer joists, sheathing, and siding are between the entire ceiling area and the weather. Cold air can penetrate here and make floors and ceilings cold.

To correct this in new construction is easy. But in an existing house, the only way to get into the area is to cut into the ceiling along the wall. Unfortunately, there is no getting around it, if you want to guard against a large heat loss.

Cut a strip along the ceiling near the wall and insert 6 inches of unbacked fiberglass along the header or stringer joists (Figure 292). Make the strip you cut as narrow as possible—just enough to slip the fiberglass into place—and make the insulation big enough to fit very snugly by friction

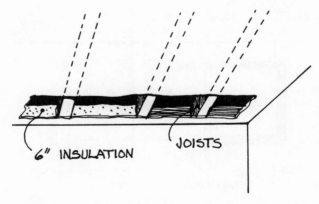

FIGURE 292. *To get to this ceiling space, remove ceiling material and apply 6-inch unbacked fiberglass to the perimeter joist, between each set of joists.*

The Hard Part

so that it doesn't flop out of position. If you build an extra inner wall, that wall will cover the opening, but you should put up a patch of plasterboard first.

If you don't install an extra wall, patch the opening by nailing plasterboard to the joists in the opening, filling the joints with joint compound and paper tape, and smoothing off with joint compound. This is easy where the joists are at right angles to the outside wall (Figure 293). It is not so easy when the opening is made where joists parallel the exterior wall (Figure 294). Here is how to do it in that case. Cut a piece of plasterboard (3/8-inch will do) 2 inches wider than the opening. Cut one inch of plaster off each side of the piece, leaving the backing paper uncut on one side to form a flange (Figure 295). Butter the edges of the opening and about an inch of ceiling along each side of the opening with joint compound; then insert the patch. The flange will adhere to the "glued" ceiling (Figure 296).

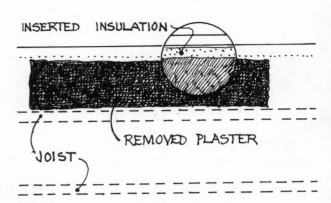

FIGURE 294. Where the removed ceiling is parallel to joists, insulation is easily installed, but the patching is difficult.

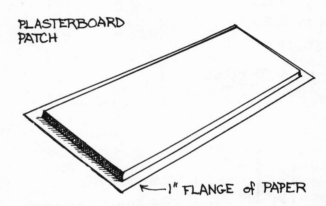

FIGURE 295. To patch plaster (or plasterboard) where there are no supports or nailing surfaces (such as between joists), make a plasterboard patch by cutting the board to leave a 1-inch flange of paper.

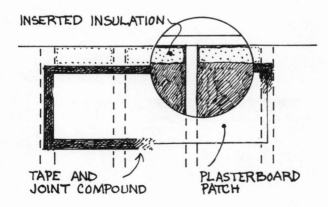

FIGURE 293. A plasterboard patch is simple to attach where the ceiling is removed across several joists, which allow a nailing surface for the patch.

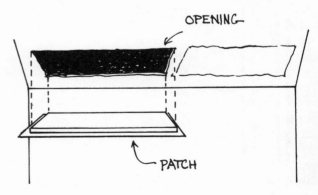

FIGURE 296. How the plasterboard patches go into the opening. Make such patches short so they are manageable and light in weight.

Smooth off the patch with joint compound and cover the edges with paper tape, finishing over the tape as you would a joint in any dry-wall system. Make these patches in medium-long increments so that each one will not be too heavy.

WINDOWS AND DOORS

Chapter 10 tells you how to handle windows and doors in new superinsulated construction. You have a similar situation in adding superinsulation to an existing house. The solutions are essentially the same, but you don't want to disturb existing windows and doors if you can help it.

One good thing about superinsulating by adding an inner wall in an existing house is that the new wall covers the weight pocket of a sash-weighted window, something that has given insulators the heebie-jeebies for a long time. Usually the solution is to cut the cords, fill the weight-pocket cavity with insulation, and put in friction-fit or spring-loaded balances. This is a pity, because there is nothing that works quite so well as a sash-weighted window. It never gets out of adjustment, works all the time, and it is easy to repair and to replace broken cords and/or missing weights. Such windows are also easy to weatherstrip.

Anyway, with the increased thickness of the wall and the casing removed, all you have to do is extend the jambs (Figure 297). The side and top stops can go in the same position; if they don't cover the joint between the existing jamb and the extender, larger ones can be installed. Windows with larger jambs can have inside storm windows, or, if there is plenty of room, self-storing shutters, as discussed in Chapter 10.

Doors are another matter. Just as in the new construction with a doubled wall, the extended

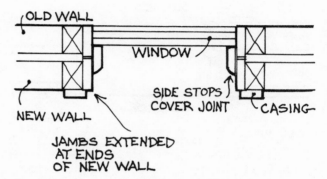

FIGURE 297. *Window jambs are extended to cover extra wall in a doubled wall in an existing house. Ordinary nominal 1-inch lumber (¾ inch thick) will do nicely. Side stops cover joint between old jamb and new. Casing finishes off the opening.*

jambs will prevent the door from opening flat against the wall (Figure 298). What can you do? You can live with the partially opening door; or you can bring the doubled wall only to the door casing, finishing off the wall end with either wood trim or wall finish (Figure 299). Or you can relocate the door to the inner part of the inner wall, which would allow it to open all the way, but it is probably not worth doing this. A fourth choice is self-evident: if the door is in a vestibule and walls are sticking out on both sides of it, there is no need to extend the inner wall.

Caulking is extremely important in superinsulation. Whether you remove the inner wall finishes of existing walls or not, there are many opportunities to caulk and seal every possible opening, joint, seam, fissure, or crack.

With door and window casings removed, caulk along the walls at corners, floors, and ceilings

The Hard Part

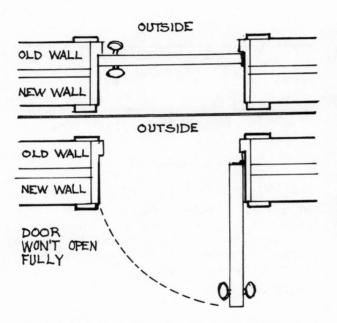

FIGURE 298. *A door poses an extra problem in a doubled wall. If door is left in its normal position (top), it will not open fully (bottom), bumping into the extra interior wall.*

(Figure 300). Use a butyl or vinyl caulk; silicone is good, too, but is expensive. When you install a new wall, either the inner stud one or just the High R Sheathing or other rigid insulation, caulk in the same places. You can run a bead of caulk on floors and ceilings and in corners when you are putting up the inner wall; this will seal the wall as it is put into place. Caulk the window casings as you reinstall them and caulk jamb extenders as you put them in place (Figure 301).

With everything done, you have a virtually airtight house. This will certainly save heat, but preventing air exchange presents other problems—all solvable.

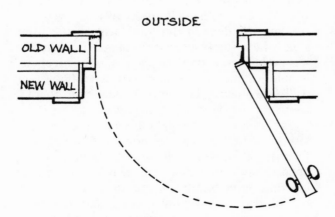

FIGURE 299. *To allow door to open a little farther than 90 degrees, the new wall is deliberately kept short of the old.*

Superinsulating Existing Walls (and Windows and Doors, Too)

FIGURE 300. *Caulk all seams and corners before installing new interior wall.*

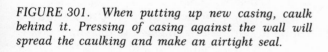

FIGURE 301. *When putting up new casing, caulk behind it. Pressing of casing against the wall will spread the caulking and make an airtight seal.*

chapter 23

Gasp!

Air Exchange and Solutions

In the "good old days," houses were built loosely, leaked air like crazy, and water vapor, smoke, odor, and other pollutants passed through the walls to the outside. So we didn't worry about them. Leaky houses also lost heat, but heat was cheap, so we didn't worry about that, either. About the only bad thing about a leaky house was the loss of water vapor in the winter—"humidity" or "moisture," as it is often called. Not only did we dry out, getting scratchy throats and stuffed-up noses, but also furniture and other wooden things dried out, too, shrank, cracked, came loose, and did other weird things.

Today, with virtually airtight houses, things are a little different. The moisture that leaked out now builds up and condenses on cool areas such as windows and even walls, where it can cause mildew and decay. Air gets stale and we get caught up in our own pollution and in gases given off by ourselves, the earth under the basement slab, and sometimes even building materials. Consider the pollutants that can build up:

• Water vapor, from cooking, showering and bathing, washing, breathing, perspiration.

• Radon, a radioactive gas usually coming out of the earth or granite rocks near a house.

• Carbon monoxide, from gas, wood, and coal stoves, and fireplaces.

• Carbon dioxide, from stoves, fireplaces, breathing, house plants.

• Formaldehyde, from plywood, particleboard, adhesives, furniture padding, and some forms of insulation.

• Chemical sprays, including insecticides.

• Dust and smoke particles.

• Any other materials in small amounts: nitrogen monoxide, nitrogen dioxide, sulfur dioxide, ozone, asbestos, lead, nitrates, and sulfates.

Water vapor, incidentally, is not necessarily a pollutant, particularly in small to moderate quantities. In fact, a relative humidity of 40 percent in the winter and at moderate outdoor temperatures (the percentage falls as temperature drops) is considered optimum for health, for warmth, and for keeping wood products in good shape. But above that percentage, water vapor is as bad as some of the other pollutants.

The solution, of course, is normally to ventilate the house by using kitchen and bathroom exhaust fans, or by opening doors and windows a few times a day. Of course, that loses precious heat. Or, some people say, don't tighten houses so much. But that gets us right back to the difficult and expensive heating system that we are trying to modify.

To obtain the best of both worlds—to relieve

Air Exchange and Solutions

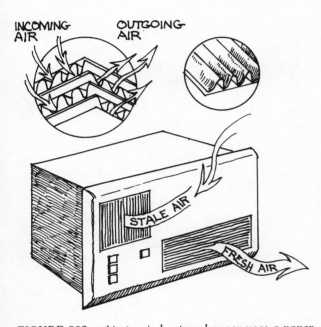

FIGURE 302. *Air-to-air heat exchanger uses a paper element (circles) to transfer heat from outgoing warm air to incoming cool air.*

houses of the buildup of pollutants without losing heat—the air-to-air exchanger was invented. It is quite new (only five years or so old), but as the industry grows, more and more companies are making heat exchangers, and their designs are slightly different from one another. Some are simply inserted in the wall, a little like an air conditioner. Others can be ducted into the house's hot-air system, or installed with their own duct system.

Basically, their function is to force out stale, moisture-saturated air from the house and bring in fresh air from outside, transferring the heat from the outgoing air to the incoming. This is done with a paper element, or one of some other material, set up in a system of tiny ducts (Figure 302). The only power used is for the motor that runs the blowers; normally the cost is not more than $60 per year.

Air-to-air exchangers do their best work in the winter in cold climates, but they also work well in hot seasons, cooling incoming air from the outgoing stale air. Most are designed to work steadily. Icing up has been one problem; this is solved by shutting the unit off for an hour out of each twenty-four hours, either manually or automatically.

One small unit may not be enough for a large house; even if its capacity were large enough, it might not exchange the air in remote rooms, or rooms that are shut off for any reason. So some are designed to work with ducts. Like stoves, heating systems, and other systems that affect the whole house, a heat exchanger must be adapted to a specific house or addition.

Prices range from $100 for a small, bathroom type to more than $1,000. The price usually depends upon the capacity of the exchanger, its air-flow volume, controls, automation, ductwork, and so on.

Gasp!

Here are some addresses of manufacturers:

ECONOFRESHER AND SENEX:
Berner International Corporation
12 Sixth Street, Woburn, MA   01801

MITSUBISHI (several exchangers):
Mitsubishi Electric Corporation
Lossnay Engineering Section, Japan
American sales company:
Mitsubishi Electric Sales America, Inc.
3030 East Victoria Street, Compton,
   CA   90221

Q-DOT (several exchangers):
Q-Dot Corporation
726 Regal Row, Dallas, TX   75247

DES CHAMPS (several exchangers):
Des Champs Laboratories, Inc.
P.O. Box 348, East Hanover, NJ   07936

ENERCON:
Enercon Industries, Ltd.
2073 Cornwall Street, Regina, Sask., Canada
   S4P 2K6
U.S. subsidiary:
Enercon of America, Inc.
2020 Circle Drive, Worthington, MN   56187

VANEE exchangers made by D.C. Heat Exchangers, Ltd., distributed by:
Conservation Energy Systems, Inc.
Box 8280, Saskatoon, Sask., Canada S7K 6C6

The Air Changer Company, Ltd.
334 King Street East, Toronto, Onto., Canada
   M5A 1K8

part IV

The Home Fires

Solar and Other Heating

chapter 24

You Are My Sunshine

Solar Devices and Collectors

Passive solar heat is simply letting the sun shine in. That sounds good, as long as the sun is shining. For the times when it isn't, there are ways of storing the sun's heat without putting in an active system; that is, one that moves air or water through tubes, ducts, and conduits. That's what you want to avoid. You want to make a simple and virtually maintenance-free system.

So it is a matter of installing collectors: a wall of windows (Figure 303); skylights and/or roof windows (Figure 304); sun spaces (Figure 305); greenhouses (Figure 306); or variations and combinations of any of these.

The collectors must face south, or nearly so, so they can absorb most of the sun's heat for the longest part of the day. Of course, south is not just south; there is true south and there is magnetic south, which can vary in the United States by as much as 21 degrees east or west. True south is the direction along any meridian to the South Pole: for example, a line from the eastern end of Lake Superior running through the center of the Florida panhandle.

A crude way to determine south on your property is to drive a stake into the ground wherever the noon sun can hit it, and observe the shadow it casts. The shadow will point north, so the opposite direction is south. And you will want to do this at solar, not clock, noon, so remember that there will be variations in solar noon, depending upon your longitude and whether you are making your determination between April and October (Daylight Saving Time). If you are, use 1:00 P.M. as your noon. If it is Standard Time, use noon.

You can be more accurate if you use a compass and a map of the United States, or even more if you have it done by an expert who might be helping you install some of your solar collecting system.

Once you have determined south, you can get an idea for the orientation of your walls, and you can build your addition, or adapt a south-facing existing wall, to collect solar heat. Slight variations should not matter in the ability of your walls and roof system to collect the heat. Ten degrees is almost nothing, and even with a deviation of 30 degrees east or west your wall will still receive up to 90 percent of the possible solar radiation. Deviation to the east is better than to the west, because the collectors will be warmed earlier on winter days and will face away from the hot afternoon sun in the summer.

And if none of your walls faces south; if, in fact, only a corner of the house does, you can build a sun space or collector that turns the corner (Figure 307), with the corner itself canted so that an angled wall does face south. There are more ways than one to get around a problem!

FIGURE 303. Wall of windows is one way to gain passive solar heat.

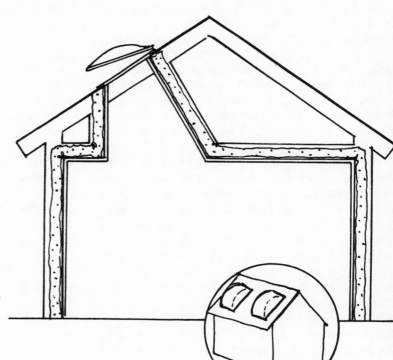

FIGURE 304. Skylight, another solar device.

Solar Devices and Collectors

FIGURE 305. A sun space is simply a glass room attached to or a part of the house, gaining solar heat when the sun shines.

FIGURE 306. A greenhouse is a sun space where plants can grow.

FIGURE 307. When an existing house does not have a wall facing south (toward the sun), sun spaces can be built in the south corner, with the angled space facing south. The other two spaces also pick up considerable sun.

You Are My Sunshine

The subject of heat storage and thermal mass is treated in the next chapter; here we will overview various collectors.

1. *Window walls*. These can be windows or sliding doors or French doors. They simply let in the sun; any masonry, such as a fireplace, nearby is part of the heat-storage system.

There must be a way of insulating these windows, not only when the sun goes down or isn't shining, so that they won't lose indoor heat, but also when the sun is hot as blazes in the summer. Shutters will help; so will the presence of deciduous trees on the south side of the house that allow the sun to shine through their bare branches in the winter and block the sun with their leafy foliage in the summer.

2. *Roof windows or skylights*. These are most effective when they open to a cathedral, or vaulted, ceiling, one that follows the contour of the roof (Figure 308). If the ceiling under a skylight is flat and the roof rises above it, a channel can be built (Figure 309) to guide light and heat into the room.

Make sure your units can be opened, to release warm air in the summer. Skylights and roof windows can be bought that are not only openable but also screened. They can be shaded, too.

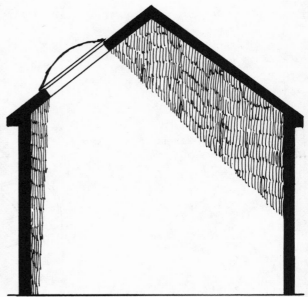

FIGURE 308

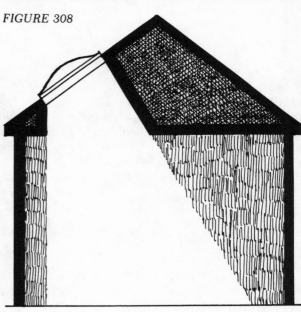

FIGURE 309

FIGURE 308. *A skylight gives most heat and light when it is installed in a cathedral, vaulted, or slanting ceiling.*

FIGURE 309. *When a skylight is over a room with a horizontal ceiling, a well or tunnel is built in the attic space to aim the light and to direct the heat, preventing it from dissipating into the attic, where it is neither desired nor useful.*

3. *Sun space or greenhouse.* A sun space is a fancy name for a greenhouse but may have less glazing than a greenhouse. We will treat both of these at once. It is an addition on the south side of the house that simply collects heat and allows it to come into the living space. It should be vented, and the vents should be such that they can be closed when the sun goes down. Fans are a good idea in these spaces, because they can force the warm air into the house and they can be shut off automatically when the air cools too much in the sun space.

Sometimes a greenhouse with a lot of plants growing in it should not be allowed to cool too much at night in the winter; in this case the fans can reverse themselves and bring warm air from the house into the greenhouse. Greenhouses and sun spaces also should have vents to the outside, to prevent the buildup of too much heat, particularly in the summer.

4. *Solar porch.* This is a sun space made by converting an existing porch or sunporch to collect the heat and bring it into the house. Not only should it be vented to the outside and to the house itself, but also it should be thoroughly insulated.

5. *Solar addition.* A solar addition is one that utilizes any or all of the solar systems just discussed.

6. *Solar collector window.* A simple device, similar to an ordinary window, this consists of two window units, one on the outside and one on the inside of a wall. The outer window is double-glazed, the inner one, single-glazed. This window is designed to collect the solar heat through the outer glazing and pass it through the inner glazing into the house. Thus the inside glazing should be open at the bottom and at the top. Venetian blinds are inserted between the windows to aim the sunlight into the house (Figure 310).

The unit is also useful for effective cooling in the summer, since it allows warm air to escape through the bottom of the raised inner glazing and on through the upper opening of the outer glazing. What is happening here is a chimney effect, based on the principle that warm air rises.

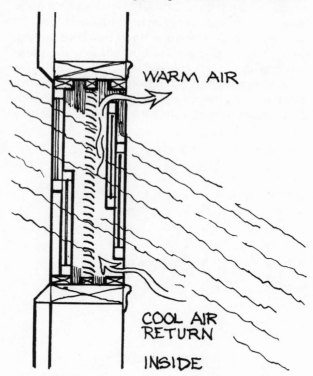

FIGURE 310. *A solar collector window is simply two windows (outside double-glazed, inside single). The inside window is open at bottom and top to allow cool air to enter the space between windows, be heated by the sun's rays, and return to the room at the top. A venetian blind helps to aim the sun's rays directly into the space.*

You Are My Sunshine

The arrangement of the openings speeds the movement of the air.

A variation is the thermosiphoning panel. Similar to a solar collector window, it has an absorber plate (not glazing) that is covered with glazing material. It is usually attached to a south wall or can be placed in a window area. Sunlight entering the panel strikes the absorber plate, heating the air inside the panel. This air rises and enters the house through an upper vent, while cool air from the house enters the panel through a bottom vent, thus setting up the convection-based, thermosiphoning effect.

Another variation is the trombé wall. It operates much like the thermosiphoning panel; it also adds the capacity to store, since it consists of a masonry wall that absorbs the heat during the day and releases it when the sun goes down.

How much glazing should be used in the various solar collectors? South-facing windows should have glazing equal to 13 percent of the floor area they will heat. That is, if the area is 2,000 square feet, the glazing should be 260 square feet, or 15 windows each about 18 square feet in size. In extra-cold climates the proportion can be increased to 15 percent; in warmer ones, reduced to 10 percent. In sun spaces and greenhouses, the glazing area should be 20 to 30 percent of the floor area to be heated.

These proportions can be varied up or down slightly. They are only guides, but you will get optimum results by sticking fairly closely to them.

chapter 25

Just So

Thermal Mass for Heat Storage

Thermal masses, properly designed and installed, can extend the heating ability of your solar collection system by a great deal. The masses store sun heat collected on a bright winter day and then release that heat to warm rooms well into the night or on cloudy days when the sun is obscured. They are, however, something of a bête-noire if not done right. If, for instance, the mass is too large, its capacity to store heat may be so great that it is unable to release it when the sun goes away. Or if it is too small, it cannot absorb all the heat available during a good day. Finally, although a mass can be in contact with the ground, it should not touch ledge or other buried rock; that simply increases the size of the mass too much.

Thermal masses should be located in the direct line of the sun or in reflected or aimed sunlight (Figure 311). They should not be located where they receive sunlight during the nonheating season. In fact, it would be ideal if the masses, or at least some of them, were situated so that cooling breezes could waft over them during the summer. They can be put almost anywhere that is practical as long as they are hit by the sun.

A mass that is on the ground should be insulated both along its perimeter and underneath. And just as a basement concrete floor or a slab on grade should have a vapor barrier under it, so should a thermal mass, to prevent water vapor from coming up through it from the ground. Remember, superinsulated houses tend to hold water vapor as well as air, so outside sources of water vapor should be eliminated or at least prevented from releasing vapor into the living space.

You can use concrete, masonry (bricks, blocks, etc.), water, stone, or phase-change materials for a thermal mass. It can be in a wall or floor, sit in a container or bin on the floor, or be mounted in a ceiling. Even a brick, concrete, or block planter or bench along a fireplace will work. The fireplace itself and its masonry wall are thermal masses, although you will get some heat loss through the roof because there is no thermal break between the masonry inside the house and outside, nor is there a practical way of putting one in.

Size, as we said, is critical to the success of a thermal mass.

A floor should be made of concrete, the darker the better, or brick, or tile. Use unglazed tile; glazed materials reflect much of the heat rather than absorbing it. For each square foot of south glazing, use (a) 2 square feet of 8-inch-thick or 4 square feet of 4-inch-thick masonry in direct sunlight; or (b) 8 square feet of 4-inch-thick masonry in indirect or diffused sunlight.

Store water in black-painted containers, or dye the water black. If you don't like black as a

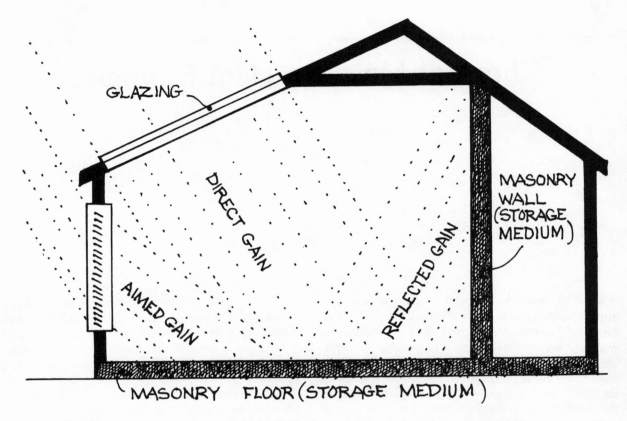

FIGURE 311. *Direct or reflected solar heat can be aimed, as by a venetian blind, to a storage medium such as masonry. Another good storage medium is black barrels full of black (dyed) water.*

container color, use brown or maroon or some other dark color. Use 6 to 9 gallons of water in direct sunlight for each square foot of south glazing.

Rock, large gravel, or crushed stone (1½ to 2 inches in size) can be used in a bin on a floor. Use one cubic foot of rock or stone for each 2 square feet of glazing or surface that is aiming heat to the rock.

Although phase-change materials are not new, having been invented or discovered in the 1930s and called Glauber salts, or eutectic salt solutions, they are new in solar-collecting applications. They change from a solid or semisolid state to a liquid state when they are heated by the sun (or any other heat source). In so doing, the materials store a great deal of heat and hold it until the air cools. When they cool, they slowly release that heat as they return to the solid or semisolid state. Several types are available, including rods, pods, pads, and quart-sized cans. Use 1 to 1½ gallons of phase-change materials in direct sunlight for each square foot of south glazing.

Thermal masses should not be covered. Wall-to-wall carpeting or a wood floor over a masonry floor would defeat the purpose. You can use area rugs, but try to leave 30 percent or more of the masonry exposed to sunlight. The same goes for walls; avoid tapestries, draperies, and the like. Pictures are okay.

chapter 26

Warmth Plus

Auxiliary Heat

With a superinsulated addition, possibly with a superinsulated house, and with some sort of solar heat, as well as an existing heating system, you may be feeling *too* warm. Auxiliary heat may seem quite unnecessary. On the other hand, you may find that using your present heating system for the superinsulated building or addition is inefficient; that is, so little heat is actually needed to bring the temperature into comfort range (68-72 degrees) that it is just too costly. Many houses built before insulation was installed as a customary thing have oversized heating systems, designed to keep warm a house that leaked air and heat like a sieve.

So, auxiliary heat may be the ticket: small enough to be economical, big enough to heat an addition; or even, with superinsulation and solar heat, a whole house. There are several ways to do it.

Baseboard radiators using hot water may be good. These can be installed in minimal lengths and can use the hot water provided by the domestic water system rather than require a boiler of their own. Oil or gas is the best fuel to use for such a system.

A wood or coal stove is another type of auxiliary heat. It requires a chimney, which can be masonry and thus work as a thermal mass. If the chimney is mainly inside rather than outside of the house, this will prevent heat loss. An uninsulated chimney (a metal stack) in the occupied part of the building also radiates heat; but you must be sure this is legal. An 85,000-Btu stove can heat 2,000 square feet of floor area. If the living space is superinsulated, the stove can be much smaller. A stove should be centrally located, and the area to be heated should be kept as open as possible to allow the free circulation of warmed air (Figure 312).

The important things concerning a stove are its efficiency and its safety: these criteria can both be met with airtight wood-burning and coal-burning stoves. Safe insulation for both stove and chimney is essential, too. Proper material must be used for the chimney, and all flammable surfaces near the stove must be protected or the stove set sufficiently far from them. Each state has regulations about stoves and instructions as to their installation. Usually a permit is required. A stove next to a masonry wall or on a masonry floor is good because the masonry will absorb heat while the stove is working and will release it when the stove is off—a thermal mass, in short.

Fans are an important adjunct to a superinsulated, solar- or auxiliary-heated house. Ceiling fans that are reversible can bring hot air downward during a cold season, and, reversed, bring cool air upward during a hot period (Figure 313).

Warmth Plus

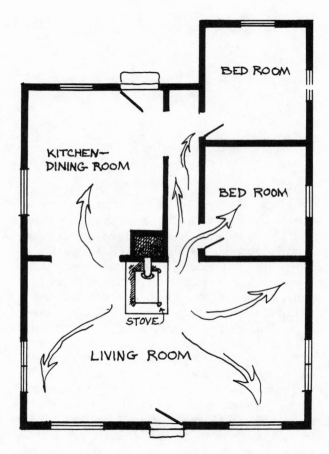

FIGURE 312. *A stove must have a central location to heat an entire floor, and it helps if the floor plan is open, or doors kept open or ajar to aim the heat flow.*

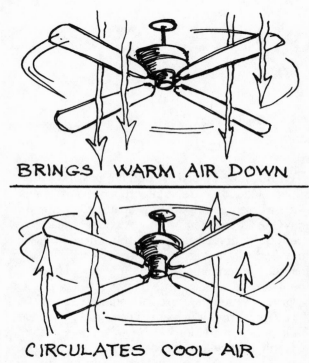

FIGURE 313. *An old-fashioned ceiling fan brings warm air from the ceiling in winter and is reversed to bring cool air up and keep it moving in the summer.*

Auxiliary Heat

A ceiling fan is particularly advantageous in a high or cathedral ceiling. In summertime, one fan can be aimed downward simply to get the air moving; moving air is cooling air. Or an upward-aimed fan can vent warm air out through a ceiling vent.

There are also whole-house fans, which are best installed in an attic floor. In this location they can draw air from downstairs windows and doors through the entire house and into the attic, where it is dissipated through attic vents (Figure 314). A whole-house fan will not cool the house when the outside temperature is warmer than the inside, but it certainly will when the reverse is true. And once the house air is cooled, the superinsulation will keep it cool. So fans and stoves, throwbacks to a bygone era, are not just faddish gestures to nostalgia; they can serve a real purpose in heating and cooling modern houses.

Now all you have to do is get started, and maybe you can burn your old oil or gas bills for auxiliary heat.

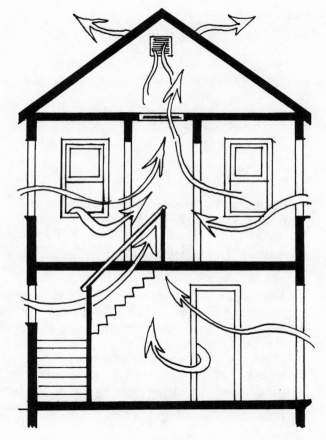

FIGURE 314. *A whole-house fan, set in the attic floor, pulls air from outside, through the house and into the attic, where it is forced outdoors through vents. A whole-house fan cools only when outside air is cooler than house air. Even when outside and inside temperatures are the same or close to each other, however, the whole-house fan gives a cooling effect by moving the air.*

Glossary

Building terms are often old-fashioned, obscure, and even quaint, but they are rarely misleading or confusing. It pays to know the terms, whether you actually build something yourself, or are in charge of the construction. It is hoped that they will be found in this glossary. There are some here that are not even in the book; and there are some nonbuilding terms that have to do with superinsulation and passive solar heating.

Abbreviations: d, penny, old English measurement for nails, still used to indicate length and girth (the original penny designation was for the price per 100 nails); o.c., on center; o.d., outside diameter or dimension; r.o., rough opening; x, for marking lumber where other pieces are to be placed, and also used for "by" as in 2 x 4.

Adhesive: Chemical or organic material used to fasten members together. Types: white, hot, contact, panel, construction, ceramic tile, resilient tile, epoxy, resorcinol, and hide.

Air exchange: A term pertaining to the air in a house that is lost and replaced by outside air. The higher the air exchange in a house (one or more times an hour, for instance), the greater the heat loss.

Air-dried lumber: Lumber that has been stored in a dry place to allow air to dry it. *See* Kiln-dried lumber.

Airway: Space between roof insulation and roof sheathing to allow passage of air; any place air can pass through.

Alligatoring: Cracking of paint that looks like an alligator's skin; caused by wear, improper application, or improper preparation of the surface. *See also* Checking.

Aluminum foil: Backing for insulation, used as a vapor barrier; sometimes used in patching. It is heavy, not the light material used in the kitchen.

Anchor bolts: Steel bolts embedded in a concrete foundation, with threaded tops to take nuts, to hold wood sills to foundations.

Apron: Trim immediately below the stool on a window. *See* Stool.

Asphalt: Residue of petroleum evaporation. Used for bituminous concrete for driveways, roof shingles, roofing felt, and roofing cement.

Attic ventilator: An opening in the attic to provide air circulation.

Glossary

Backfill: To refill an excavated area with earth, usually around the outside of a new foundation.
Baluster: Spindle of a railing, part of a balustrade.
Balustrade: Railing system on a porch or stairway, consisting of top rail, balusters, bottom rail, and sometimes a newel post.
Band: A type of molding.
Bargeboard: A decorative board covering the rake (roofline at a gable end).
Baseboard: Finish board between interior wall and floor.
Base course: First course in brick or concrete block; also called starter course.
Basement: Area below grade, usually under the first floor.
Base molding: Trim on top of a baseboard.
Base shoe: Quarter-round or similar molding nailed to a baseboard where baseboard and floor meet.
Batten: Narrow wood strip used to cover joints in vertical boards. *See* Board and Batten.
Batter board: Wood frame set into the ground as an anchor for string to determine corners of an excavation. Also used with surveyor's string to determine levelness of a foundation.
Bay: An element that protrudes from a wall, usually defined by windows. A bay window rises from the ground (foundation) one or more floors.
Bead: A strip of molding used as decoration; a strip of caulking in a joint.
Beam: Horizontal structural timber or sill supporting floor joists.
Bearing wall: Any wall that supports any load in addition to its own weight.
Bed molding: Molding at the angle between a vertical and a horizontal surface, as between eave and exterior wall. Also called cornice trim or frieze board (which see).
Bevel: To cut a board at any angle other than a right angle. *See* Chamfer.
Bird's mouth: Notch in a rafter to allow it to sit fully on the top plate of a wall.
Blind-nailing: Nailing wood so that the nailhead will now show, either through a tongued and grooved board, or through a shingle high enough so that the next shingle above covers it. *See* Face-nailing.
Blind stop: A molding just behind the outside casing of a window, sticking out enough to form a lip where the storm sash is usually placed.
Board and batten: Vertical siding made of boards with a smaller board (batten) covering the joints between them.
Board foot: Measurement of lumber. A piece of wood nominally 1 inch thick, 12 inches wide, and 12 inches long contains 1 board foot.
Boards: Lumber, nominally 1 inch thick (¾ inch actual dimension), dressed to nominal widths of 2, 3, 4, 5, 6, 8, 10, and 12 inches.
Boston ridge: Overlaid and blind-nailed shingles on the ridge of a roof.
Box beam: A beam made of 3 boards nailed together to forma *U*-shaped hollow box.
Boxed beam: A solid beam covered with boards.
Brace: A board set at an angle to stiffen a stud wall. It is a set-in brace when it is set into notches in the studs.
Brick: Clay blocks, fired for hardness and color, used to build fireplaces, walls of houses, and chimneys. Common brick is made of plain fired clay and does not resist weather very well, particularly

Glossary

when placed horizontally, as in a sidewalk. Hard brick, usually water-struck (a method of firing), is resistant to weathering.

Brick veneer: A wall one brick thick, set against a frame wall or concrete block wall.

Bridging: Wood or metal members set between joists midway in thier span to stiffen and reinforce them. Cross-bridging is boards or steel set in the form of an X; solid briding is 2-inch lumber set between joists.

Building paper: Paper used mainly as an air stop under siding, or under finish flooring. Also called sheating or resin paper.

Built-up roof: Layers of roofing felt and hot tar, topped by roofing gravel, commonly used on flat roofs. Also called tar and gravel roof.

Butt: A hinge; the hinged edge of a door; the thick end of a wood shingle or clapboard.

Butt joint: The oldest joint; where two square-edged wood members meet, end to end or at right angles to each other.

Cant strip: A triangular board set at the juncture of a flat roof and a wall to prevent cracking of applied roofing. It is like a cove (which see), to relieve the severe corner of two meeting surfaces.

Cap: Anything that tops another member; the top portion of a capital on a column, or the top piece or pieces of molding.

Capital: The top part of a column.

Carriage: A notched 2 x 12 set at an angle to carry treads and risers in a stairway. Sometimes called a stringer.

Casement window: Window hinged on one side, opening in or out.

Casng: Trim for window or door, inside or outside, nailed to the edge of jamb and wall.

Cats: Boards (2 x 4s) nailed horizontally between studs to act as nailers for wood board paneling. *See also* Fire stop.

Caulking: Pliable material, dispensed from a cartridge in a caulking gun, used to seal joints and cracks, to weatherproof, and to waterproof.

Ceiling: Any overhead surface.

Ceiling box: Octagonal junction box set in the ceiling for wiring to a ceiling fixture.

Cellar girt: A beam running from front sill to back sill, alongside a chimney, in a post and beam house.

Cement: A baked, powdered stone (often called portland cement) mixed with sand, gravel, and water to make concrete.

Cesspool: A hole in the ground, sometimes lined with stone, into which plumbing wastes are directed, to act as a septic tank in breaking down sewage into solids (sludge) and liquids, which drain away through openings in the hole and dissipate in the ground. A cesspool is illegal in most communities. *See also* Septic tank.

Chalk line: A string saturated with chalk, used to mark a line from one point to another on surfaces. It is normally held taut and snapped to make its mark.

Chamfer: A beveled edge.

Checking: Cracks in paint.

Glossary

Check rails: In double-hung windows, the bottom of the top of the sash and the top of the bottom sash, meeting in a weathertight joint.

Checks: Small splits or cracks in wood, due to improper seasoning or curing.

Chimney: A vertical tube to remove smoke and gases from fires, furnaces, and stoves.

Circuit: A part of the electrical system, designed to carry a limited amount of electricity.

Circuit breaker: A device for interrupting the current in an electrical circuit when there is a short circuit or overloading. A modern substitute for a fuse.

Clapboards: Beveled boards overlapping horizontally, sometimes shiplapped, used as siding; made of cedar or redwood.

Cleat: A strip of wood attached to another to hold a third in place.

Clinch: To bend over the point of a nail driven into two pieces of wood; a very strong fastening.

Collar beams: Nominal 1- or 2-inch boards connecting opposite roof rafters to prevent them from spreading. Also called collar ties.

Collector: Anything that collects the sun's heat in a solar heating system. It can be ordinary glazing (windows), skylights, roof windows, or special heat-absorbing plates.

Column: Vertical supporting member; post; pillar. Decorative columns are tapered from bottom to top, sometimes even bellied to appear tapered. When a column is made of concrete and is large, it is called a pier.

Combination door, window: Serving as both storm sash and screen, generally self-storing.

Concrete: Mixture of cement, sand, gravel, and water, very strong when dry. Reinforced concrete has added steel mesh or rods.

Condensation: Water formed from water vapor hitting a cold surface. It is to be avoided in houses, on windows, and in crawl spaces, attics, and basements.

Conduit: Metal tube containing electrical wires.

Cope, or coping: Method of forming the end of a molding to follow the face of adjacent molding in an inside corner. It is done when ordinary mitering will not fit a corner that is not 90 degrees. The end of the molding is mitered; then the contour of the inside edge of that miter is cut at right angles to the length of the molding with a coping saw (one with a small narrow blade). The resulting form will fit the facing of the adjacent molding.

Corbel: To offset brick so that it angles in or out. Used to increase the size of a chimney or to change its direction. Also used, more rarely, to change the direction of a wall.

Corner: Place where two perpendicular walls meet. An inside corner has a 90-degree angle; an outside corner has a 270-degree angle.

Corner bead: An *L*-shaped wood member that fits an outside corner for protection and decoration; an *L*-shaped steel strip nailed to an outside corner and covered with plaster or joint compound for a finished corner.

Cornerboard: An *L*-shaped strip formed by two boards and nailed to the outside corner of the exterior of a house, to which siding is butted.

Cornice: Boxed structure at the eave (overhang) of a roof, consisting of facia (face of the eave) and soffit (underpart of the overhang). Sometimes includes frieze board (which see).

Glossary

Cornice return: The part of a cornice that turns a corner, from the eave to the bottom of the rake (which see).

Counterflashing: Metal flashing mortared into a chimney and covering step flashing where a slanting roof meets the side of a chimney.

Countersink: To recess a nailhead into wood; to make the top of a screw hole larger than the shank hole in order to recess its head.

Course: Any horizontal row of bricks, blocks, clapboards, or shingles.

Cove: Concave wood molding used in interior corners, where wall and ceiling meet; any concave surface filling the area where a horizontal surface meets a vertical surface.

Crawl space: Area between first floor and ground, enclosed or open, but not high enough to act as a basement.

Cricket: See Saddle. Also, the term for a triangular piece of nominal 2-inch lumber nailed to an unnotched stair carriage onto which treads and risers are nailed.

Cripple studs: Short studs above and below windows and above doors.

Crosscutting: Sawing wood across the grain. The opposite of ripping (which see).

Curing: Allowing lumber to dry, or plaster or concrete to set and dry.

Cutting in: Painting a surface up to a corner, without getting paint on the other surface. Done with a steady hand and skill and a good-quality brush; avoids using masking tape or a paint guide.

Dado: A groove cut across the width of a board.

Decay: Rotting of wood due to moisture and/or fungus.

Dentil: A rectangular block forming one of a series, used as an ornament on molding. Dentils look like square teeth.

Dimension lumber: Boards nominally 2 to 5 inches thick.

Dimple: To depress nailheads in plasterboard without breaking plasterboard paper, so they can be covered with joint compound.

Direct nailing: Nailing through one member into another, with the nailhead showing. Also called face-nailing.

Door: A member designed to close an opening, including everything but windows.

Dormer: Roofed structure covering an opening in a sloping roof. A shed dormer has a sloping roof on one plane, and is designed to add more space under a roof. "A," or eye, dormers are primarily for light and ventilation.

Double-hung window: Two sashes in vertical grooves that bypass each other when raised or lowered; a vertical sliding sash.

Downspout: Vertical wood or metal tube to direct water from gutters to ground, storm sewer, or dry well. Also called leader.

Dressed lumber: Lumber planed down from its rough (full) size to its dressed nominal size. A rough 2 x 4 is a full 2 by 4 inches. A dressed 2 x 4 is 1½ by 3½ inches.

Drip: Projection at the edge of a roof to allow water to drip over the edge rather than run down the face of an exterior wall; a groove on the underside of a drip cap or windowsill that prevents water from

following the contour of the wood and dripping down the wall. The groove interrupts the flow of water.
  Drip cap: A wood molding set on top of a window or door casing to divert water.
  Drip course: Projecting course of masonry to deflect rainwater.
  Drip edge: Metal strip in the shape of an *L* to extend roofline and act as a drip.
  Dry wall: Plasterboard as an interior wall covering. Sometimes skimcoated with plaster.
  Dry well: A hole in the ground, sometimes filled with large stones or supersized gravel, to allow for drainage of water from gutters or such things as laundry washers into the ground.
  Ducts: Round or rectangular pipes for distributing warm air from a furnace. Made of galvanized steel.
  Duct tape: A silver, adhesive tape used originally to seal joints in heating ducts; now used to seal insulation, vapor barriers, and joints in virtually anything.
  Duplex receptacle: A place to plug in two electric plugs.

  Eave: Edge of roof along the lowest part of the roof.
  Eaves trough: See Gutter.
  Elevation: Front, side, or back view of a structure, in a drawing.
  Entasis: A marvelous word meaning the slight convex curve in the taper of a column to make the sides appear straight.
  Excavation: A hole in the ground for cellar, foundation, footings, pipes, and cables.
  Expansion joint: Asphalt-impregnated fiber strip placed in full-depth joints in concrete to prevent the concrete from cracking.
  Expansion plug: A fiber, plastic, or lead sheath that, when inserted in a screw hole, expands as the screw is driven and holds the screw fast. Usually used in masonry.
  Exterior plywood: Plywood made with waterproof glue.

  Face-nailing: Direct nailing through one member into another. Nailheads are exposed.
  Facia: The face, or front, of a cornice or eave; horizontal board just below the roofline.
  Feather edge: A board trimmed to a thin edge to fit into the groove of another board; used in early paneling. Also (as a verb) to sand or plane the edge of paint or other material down to a thin edge.
  Felt: Paper saturated with asphalt, sometimes called tar paper. Used sometimes under roof shingles, siding, and finish floors.
  Fenestration: Windows.
  Fieldstone: Natural stone used in foundations and retaining walls, sometimes in house walls. Fieldstone is usually not cut or broken.
  Filler: Anything used to fill holes or gaps in wood, and sometimes in plaster. It can be wood or putty; distinguished from wood filler, which is a thin, pastelike material used to fill pores in open-pored wood such as mahogany, oak, and walnut.
  Finish: Any covering, such as paint, stain, varnish, wallpaper; a covering for a wall, such as plaster or plasterboard; final work, such as finish carpentry as opposed to rough carpentry.
  Fireback: Steel or iron sheet placed at the rear of a fireplace to reflect heat into a room.

Glossary

Fire stop: Horizontal 2 x 4 or similar board set between studs to retard the spread of fire in a hollow wall.

Fishplate: A board or plywood connecting beams or rafters butting end to end; sometimes applied to a ridgeboard between rafters, meeting at roof ridge. Also, a steel beam set between two wood beams.

Flagstone: Flat stones, usually cut, used for floors and sidewalks, sometimes for retaining walls.

Flashing: Metal or roll roofing placed between roof and wall, to make joint weatherproof; can also be placed in a roof valley where two slanting roofs meet.

Flat: Not shiny.

Floor: The base of each story of a house; any surface that can be walked on, other than steps.

Flooring: Any finish surface for a floor.

Flue: Passage in a chimney for fumes and smoke.

Flue lining: Clay tubing made in short lengths, and metal in long lengths, to fit inside a chimney to keep brick and mortar from deteriorating from heat and gases.

Fly rafter: End rafter of a gabled roof, overhanging the gable wall and supported by lookouts and roof sheathing.

Footing: Concrete platform, wider than a foundation, on which the foundation sits. It should be below frost line to prevent the foundation from heaving because of freezing and thawing. Also placed under piers and posts.

Forms: Wooden members made of plywood and 2 x 4s, used to hold concrete until it sets.

Foundation: Wall of concrete or concrete block set in the ground. It holds up the house.

Framed overhang: A floor overhanging the floor below it, with joists extending beyond the exterior walls supporting it. A garrison-style house is an example.

Framing, balloon: Also called American light wood framing; an early system using 2-inch lumber in which studs extend from sill to roofline and on which floor joists are hung.

Framing, platform: Also called western wood framing; a system of wood framing in which each floor is built separately as a platform for the walls.

Framing, post and beam: Also called brace framing; a system of wood framing using heavy beams and posts, set on wide centers, with slightly smaller joists and studs.

Frieze or *frieze board:* Horizontal board connecting the soffit of a cornice to the siding.

Frost line: Depth to which earth freezes in winter. Footings for a foundation must be below the frost line to prevent heaving from freezing and thawing, and other movement of the structure.

Fungi: Plants that grow in damp wood, causing mildew and decay.

Fungicide: A chemical that kills fungi.

Furring, or furring strips: Strips of wood, usually 1 x 2 or 1 x 3, sometimes of metal, applied to studs or joists, sometimes directly to the wall, to even out a rough wall or ceiling; a base for securing a finish wall or ceiling, such as plasterboard, plywood paneling, boards, or lath. Also called strapping.

Fuse: A device to protect electrical apparatus and wires against excessive current. It has generally been replaced by the circuit breaker.

Gable: Triangular part of an exterior wall created by the angle of a pitched roof.

Galvanizing: Coating metal with a layer of zinc to inhibit rust.
Gambrel roof: A double-pitched roof with steep lower pitch and shallow upper pitch, characteristic of Dutch Colonial design.
Girder: A heavy beam.
Glazing: Glass and the installation of glass in windows and doors.
Glazing compound: A modern putty to waterproof panes of glass in a wood window sash.
Grade: Surface of the ground.
Grain: Direction of fibers in wood.
Greenhouse: A glazed building used to grow plants in cool and cold weather. It usually needs auxiliary heat when the sun is not shining.
Groove: A notch running the length of a board.
Ground cover: Sheet or roll material (plastic or asphalt paper or felt) to cover the ground in a crawl space or cellar, to prevent moisture from rising from the ground. Also, a low-growing plant.
Grounds: Wood strips (1 x 1s, generally) nailed around door and window casings and next to floor and ceiling, used as guides for the correct thickness of plaster.
Grout: Mortar thinned to a runny consistency and virtually poured into masonry joints; mortar designed to fill the joints in ceramic tile.
Gussets: Plywood or other wood connecting joints in wood, usually connecting individual members of a roof truss. *See* Fishplate.
Gutter: Channel of wood or metal to drain water off the roof into downspouts. Also called eaves trough.
Gypsum board: Plasterboard.
Gypsum plaster: Most common kind of plaster.

Half-lap: Two pieces of lumber cut with dadoes to half their thicknesses where they cross or meet on a corner so that they fit together in the thickness of one piece.
Hardboard: Manufactured sheet material made of ground-up wood fiber, used mainly as an underlayment for tiles and sheet flooring. Comes ⅛ to ¼ inch thick.
Hardware cloth: Heavy steel mesh, with ¼- to ½-inch holes.
Header: A beam placed at right angles to the ends of floor joists to form openings for chimney, stairway, fireplace, etc.; a beam placed as a lintel over door and window openings; a brick laid so that its short face, or head, shows in the wall, and used to connect a second layer of bricks in a double-thick wall.
Header joist: Floor joist connecting the ends of regular floor joists and forming part of the perimeter of the floor framing. Opposite of stringer joist. Both are called ribbon joists.
Hearth: Floor of a fireplace.
Heat storage: A system for holding heat from the sun. Can be masonry, water, crushed stone, or gravel. Heat is released when the source of the heat is removed.
Hip: Outward-sloping ridge formed when two sides of a roof meet. Opposite of valley.
Hip rafter: Joist that forms the hip of a roof.
Hip roof: A roof that slopes up from all four sides of a structure. Can come to a peak, but usually has a short ridge.

Glossary

Homasote: A fiberboard sheet made of ground-up newspapers, virtually a papier-mâché. Used as a ceiling material. Modern Homasote has a special covering and has many uses.

I-beam: A steel beam, named for its profile shape, used to support joists in long spans, and as an extra-long header over windows and doors.
Insulation: Any material used to retain interior heat.
Interior finish: Material covering interior wall frames: plaster, plasterboard, or wood.
Intrinsic heat: Any heat that is not manufactured from a fuel or the sun; usually applies to the heat in a house from breathing, perspiring, cooking, washing, or from lights and appliances.

Jack rafter: A short rafter spanning the space from a hip to the top plate of a wall, or from a valley to the roof ridge.
Jamb: Side and top frame of a window or door, against which the window or door fits. The top jamb is called a head jamb.
Joint: Any space between two components.
Joint compound: A plasterlike material to cover nailheads or screwheads and joints in plasterboard wall construction. Joints are also covered with paper tape.
Joist hanger: A metal fastener securing the end of a joist directly against the side of a girder or other joist. Larger hangers, made for larger beams, are called timber supports or beam hangers.
Joists: Nominal 2-inch-thick horizontal beams, set parallel to support a floor or ceiling. A floor joist is set on sill and beams; a ceiling joist is set on the top plates of walls.
Junction box: An electrical utility box used to house a spliced electric wire.

Kerf: Cut made by a saw.
Kiln-dried lumber: Lumber dried in an oven or kiln to reduce its moisture content. *See* Air-dried lumber.
Knee wall: A short wall, usually connecting floor and slanting roof.
Knot: In lumber, a round spot of wood harder than its surroundings, the result of cutting a log where a branch has grown.
Knothole: Place where a knot used to be.

Landing: A platform dividing a flight of stairs into two sections.
Lap joint: A joint in which one member of a doubled beam or plate overlaps the other member. The most common lap joints are found at the corners where wall top plates, made of doubled 2 x 4s, meet. Lap joints can also be cut out of solid wood components, such as 4 x 6 sills.
Lath: Base for plaster, made of wood, metal, or plasterboard.
Lattice: Framework of crossed wood or metal strips; a board, usually ¼ inch thick and 1½ to 2 inches wide. Commonly used to skirt open spaces under porches.
Leader: A downspout.

Glossary

Ledger: Heavy strip of lumber nailed to girder, joist, or wall, onto which other joists or components are set.

Let in: To notch a series of studs to receive a board so that it is flush with the stud surface. A let-in brace is such a board.

Level: Horizontal; perpendicular to vertical, or plumb.

Light: One pane of glass in a window, named for its ability to admit light.

Linseed oil: An oil made from flax, used in paints and to condition and finish wood.

Lintel: Horizontal component supporting the opening above a door or window. Also called header.

Lookout: In a roof overhang, a short horizontal bracket connecting rafter end to wall, covered by facia and soffit.

Lookout joist: Horizontal joist overhanging an exterior wall and cantilevered over that wall, usually employed in window construction or when the second floor of a house overhangs the first-floor wall.

Louver: An opening with angled slats to keep out the weather, and screened to keep out insects and vermin, which allows entry and exit of air. Used to ventilate attics and crawl spaces.

Lumber: See Boards. Dimension lumber; Dressed lumber; Matched lumber.

Lumber core: The thick interior wood between two thinner pieces in lumber-core plywood. Most plywood is made up of an odd number of equally thick pieces of wood and is called veneer plywood.

Mantel: Shelf above a fireplace, including trim around the fireplace opening. The shelf can be wood or masonry.

Masonry: Stone, brick, or concrete block held together with mortar.

Matched lumber: Boards with a tongue in one edge and a groove in the other, designed to make a strong, tight joint. End-matched lumber has a tongue or groove in one end of each board.

Millwork: Lumber shaped or molded in a millwork plant.

Mineral spirits: A petroleum solvent used as a substitute for turpentine. Also called paint thinner.

Miter joint: A joint made with two pieces of lumber (or any other components) each cut at a 45-degree angle to form a 90-degree corner.

Molding: Decorative wood strips and boards, used as interior and exterior trim.

Mortar: Material used to hold masonry components together. Made of cement, lime, and sand, and enough water to hold it together.

Mortise: a rectangular or square hole cut in wood to receive a matching tenon, or tongue, of another component, to make a mortise and tenon joint. The mortise is the female part of the joint. *See* Tenon.

Movable sash: A window that opens and closes.

Mullion: Vertical divider between two windows and/or door openings.

Muntin: Part of a window sash frame dividing lights of glass.

Nailer or nailing block: A wood member attached to a surface to provide a nailing surface for attaching another member.

Nails: Metal fasteners designed to hold one component to another. Types include *common,* used for general framing and fastening; *box,* thin-shanked for finer work; *casing,* with medium-sized heads for

Glossary

countersinking on exterior casings; *finishing*, with small heads for countersinking on interior work; *cut*, for hardwood flooring; *spiral*, or *screw*, for special holding qualities; *masonry*, for driving into concrete and other masonry components; *roofing*, large-headed, for holding down asphalt shingles; *shingle*, for nailing wood shingles; and *ring-shanked*, with rings instead of spirals, for underlayment and plasterboard. Most nails are galvanized; hot-dipped zinc galvanized nails are the best of the galvies.

Natural finish: A finish designed to show the grain and color of wood. Usually varnished, shellacked, or lacquered. *See* Stained finish; Painted finish.

Newel or newel post: A post to which a railing or balustrade is attached.

Nominal: See Dressed lumber.

Nonbearing wall: A wall supporting no load other than its own weight. *See* Bearing wall.

Nosing: Any projecting edge of molding, particularly the projecting part of a tread, over a riser, in stairs.

Notch: Cross-grain rabbet at the end of a board.

Oriel window: A bay window that projects from the wall and is carried by brackets, corbels, or a cantilever, usually on a second-floor wall. It is often confused with a bay window. The difference is that a bay starts at the ground; an oriel begins above the first floor.

Outlet: An electrical receptacle or box mounted in or on a wall and connected to a power supply, used to hold a socket for a plug.

Painted finish: Wood or other surface covered with an opaque, pigmented finish of any color, called paint. *See* Natural finish; Stained finish.

Paint thinner: Any solvent that reduces oil-based paints; mineral spirits.

Palladian window: A triple window, with the center taller than its flankers.

Panel: A thin piece of wood fitted into grooves in the stiles and rails of a door.

Paneling: Any wood wall finish.

Paper: Term for papers or felt applied under finish floors, siding and roofing. Also called sheathing paper and building paper.

Parquet: Geometrically patterned floor of wood tiles or boards.

Particle board: A sheet made by gluing wood chips or particles together under pressure. Used as an underlayment for resilient tiles and carpeting.

Parting bead: A thin strip of wood inserted in a window jamb to act as a divider between upper and lower sashes of a double-hung window.

Partition: A wall that subdivides space in a building.

Pediment: A triangular gable over a window or door. Variations include broken and scroll pediments.

Pendill: Carved wood drop at the lower ends of a second-floor overhang. Also called pendant.

Perspective: A drawing representing what the viewer actually sees.

Phase change: Refers to the change of certain materials from a solid or semisolid state, when they heat up, into a liquid state, when they store the heat. When the heat source is removed, the material reverts to the solid or semisolid state, releasing its heat.

Pier: Column of concrete or heavy masonry.
Pie steps: Wedge-shaped treads in a stairway turning a corner. Also called winders.
Pigment: Opaque coloring in paint or stain.
Pilot hole: Hole drilled in wood to receive a screw or nail.
Pitch: Slope of a roof.
Plan: Drawing of a building (floor) as seen from above, with the roof removed.
Plank flooring: Any wide boards used for flooring.
Plastic: Materials such as urethane, polystyrene, polyethylene, polyvinyl chloride, vinyl, and related compounds, used as building and plumbing materials.
Plate: The top and bottom of a stud wall. The floor plate, also called sole plate, is the bottom horizontal member of the wall, usually a 2 x 4 or 2 x 6. The top plate is the top horizontal member, doubled, supporting second-floor joists or roof rafters. A sill plate is also called a sill, a wood member sitting on the foundation and supporting floor joists.
Plow, or plough: A groove along the face of a board, not on an edge. Also called simply a groove.
Plumb: Vertical; perpendicular to horizontal, or level.
Plywood: Sheet wood made by laminating thin pieces, or plies, together, each ply with the grain running perpendicular to that of the next ply. Will not split as ordinary boards might, and is very strong for its weight.
Polyethylene: A semiclear plastic sheet material used as a vapor barrier.
Polyisocyanurate: A plastic foam, related to urethane, used as rigid insulation.
Polystyrene: A plastic in foam form, used as rigid insulation.
Preservative: Fluid, with a copper, zinc, or pentachlorophenol base, to prevent or retard decay in wood. Modern pressure-treated wood has preservative in a gaseous state forced into the heart of the wood under pressure.
Primer: First coat of exterior paint jobs of more than one coat. Interior primer is called enamel undercoater or enamel undercoat.
Putty: A powdered material that is mixed with water and used to fill nailheads and wood cracks; an obsolete word for glazing compound. Putty can be ready-mixed and colored.

Quarter-round: A molding that is one-quarter of a dowel. Half-round is a half dowel and full round is a full dowel.
Quoin: Stone or masonry block forming an outside exterior corner; in wood construction, wood members forming an outside exterior corner to simulate masonry blocks.

R factor: Resistance of a material to the loss of heat from one area to another. The higher the R factor, the greater the resistance.
Rabbet: A notch running the length of a board, on the edge of the board.
Rack: To be forced out of plumb; when a wall or building goes from a rectangle to a parallelogram.
Rafter: Sloping joist, used to hold up a roof. Flat roof rafters are sometimes called roof joists. *See also* Hip rafter; Jack rafter; Valley rafter.

Glossary

Rail: Horizontal member of a window or paneled door; upper or lower horizontal member of a balustrade.
Rake: Gable end of a roof from eave to ridge; trim board along the roof slope to finish off the edge.
Rebar: Reinforcing rod or bar.
Reinforcing: Steel rods or mesh placed in concrete to strengthen it.
Relative humidity: The amount of water vapor in the air. The warmer the air, the more water vapor it can hold.
Retaining wall: A masonry wall designed to retain earth behind it.
Ridge, ridgeboard, ridgepole: Horizontal member, nominally 1 or 2 inches thick, forming the ridge of a roof where the tops of the rafters meet. Sometimes called rooftree.
Ripping: Sawing a board with the grain.
Rise: In stairs, the vertical height of a flight of stairs or the height of one step; in roofs, the vertical height of a roof from wall plate to the ridge.
Riser: Board enclosing the space between treads of a stairway; or the space itself. Steps without riser boards have open risers.
Roof: Sloped or flat surface covering the top of a building.
Roofing: Any material on the roof to keep out the weather: shingles (wood, asphalt, slate, etc.), metal, roll roofing (asphalt-saturated felt and roofing gravel).
Roofing felt: Asphalt-impregnated paper used under roof shingles and certain flooring and siding materials. Also called tar paper.
Rough framing: Bare framing with wood members, including sheathing.
Rough opening: Opening in a wall for a door or window; opening in a floor for stairway, chimney, or fireplace.
Rubble: Rough, broken stone, block, brick, or concrete, used as a filler material.
Run: In stairs, the horizontal length of a stairway; in roofs, the horizontal or level distance over which one rafter runs; half the span of a double-sloped roof.

Saddle: A double-sloped structure installed uproof of a chimney to prevent buildup of snow or rain against the chimney. Also called a cricket.
Sash: A single window frame containing one or more lights of glass.
Sash balance: A spring or weight designed to hold a window open, closed, or anywhere in between.
Satin finish: Semigloss or less than semigloss finish of paint or varnish.
Screed: A board used to level fresh concrete; as a verb, to scrape a board across concrete to level it in forms after it is poured.
Screw: Spiral-shanked fastener for wood and metal, turned as it is driven; needs pilot hole.
Scribing: Fitting woodwork or paneling to an irregular surface; using a scribe (compass) to transfer an irregular surface to woodwork or paneling that is then cut to fit.
Sealer: A liquid designed to seal the surface of wood as a base for paint, varnish, more sealer, or wax.
Section: A drawing representing a side view of a house with walls removed.

Glossary

Septic tank: A domestic sewage-disposal system, consisting of an enclosed tank buried in the ground and connected to a house sewer system. Solids collected in the tank are broken down by bacteria; liquids are siphoned off into a leaching field, where they are absorbed into the ground. *See* Cesspool.

Shake: A thick wooden shingle, usually split but sometimes split on one side and sawn on the other. Used for wood roofs and rustic siding.

Sheathing: Exterior covering of a wall, a base for siding; exterior covering of a roof, a base for roofing.

Sheet flooring: Any finish flooring material that comes in sheets, as opposed to strip or tile flooring. Vinyl flooring and linoleum are examples.

Sheet-metal work: Nearly anything made of sheet metal, such as ducts in a hot-air heating system, gutters, downspouts.

Shims: Tapered pieces of wood, generally shingles, used to close gaps between members.

Shingles: Small pieces of building material used for siding or roofing. Siding shingles are wood, sawn and tapered, made from red or white cedar. Roofing shingles are made of asphalt, fiberglass, metal, wood, slate, etc.

Shiplap: A rabbet along the side of a board, to allow a board to overlap another with their surfaces remaining on the same plane.

Shutter: Hinged exterior covering for a window, usually folded back against the wall. Originally for protection against weather, later only for decoration, now coming back into favor as a protector. Inside shutters are generally used to reduce heat loss.

Siding: Exterior covering of a wall to keep the weather out and to "clothe" a building.

Sill: Timber sitting directly on the foundation, the support for floor joists; in windows, the slanting exterior bottom piece of a window frame.

Sill sealer: Fiberglass strip inserted between foundation and sill to seal any variations in the foundation and to keep weather and cold out.

Skimcoat: Any thin (usually ⅛- to ¼-inch) coating, such as plaster, stucco, concrete, or mortar, on a surface.

Sleeper: Board, nominally 2 inches thick, secured to a concrete floor as a base for a wood floor; a 2-inch-thick board connecting two ceiling joists to act as a nailer for a stud wall paralleling the joists.

Soffit: The underside of an eave overhang.

Soil stack: Vent pipe for plumbing and main drain for house sewage. The same pipe serves both functions.

Solar heat: Heat from the sun. A passive solar heating system lets the sun shine in and uses no moving parts. An active solar system heats air or water from the sun and circulates it by fans or pumps to the area where it is needed. A passive solar system can use fans to move heated air; in that case it is usually considered modified passive or semiactive.

Spackling compound: A pliable, plasterlike material to fill narrow cracks and holes in plaster.

Span: Distance between supporting points. In a roof, the total level distance between rafter supports.

Spline: Also called loose tongue. A thin strip of wood placed in grooves in the edges of adjoining boards to form a joint.

Square: One hundred square feet, a unit of measurement for roofing, and sometimes for siding. The

Glossary

100 square feet is that area actually exposed to the weather, not the entire area of the uninstalled material. Also a tool (T-square, framing square) used for marking right angles and other measurements.

Stained finish: Colored or pigmented stain applied to a wood surface and varnished, shellacked, or lacquered. *See* Natural finish; Painted finish.

Stair: Steps leading up and down.

Stair carriage. See Carriage.

Stile: Vertical piece in a paneled door or window sash.

Stool: Interior molding fitted over a windowsill, erroneously called a sill.

Story: Living area between floor and ceiling on one level of a building.

Straightedge: Anything straight, used to check for level and straight surfaces.

Strapping: See Furring.

Stretcher: A brick laid lengthwise in a wall.

Stringer: See Carriage.

Stringer joist: The border joist of a floor frame, parallel to intermediate joists. Opposite of header joist.

Strip flooring: Narrow wood floorboards. *See also* Plank flooring.

Stucco: Siding made with cement-based plaster, applied over metal lath.

Stud: Vertical member in a wood frame wall. Sometimes made of metal.

Subfloor: Rough boards or plywood secured to floor joists, onto which a finish floor or underlayment is secured.

Sun space: A glazed building or structure designed to collect the sun's heat and disperse it into a building. Can use storage techniques. Needs no auxiliary heat.

Superinsulation: A term used for extra-thick insulation in a building's walls, attic floor, and basement ceiling. To be superinsulated, a unit must have an R factor of at least 30 for walls, 40 to 60 for attic floors, and 19 for basement ceiling.

Suspended ceiling: A ceiling hung from joists by brackets or wires; a ceiling not secured directly to joists or furring strips.

Tail beam: A short beam supported on one end in a wall and on the other by a header.

Tenon: Projection of a stud or other member cut to fit into a regular hole, or mortise. The tenon is a male part; the mortise the female part. *See* Mortise.

Termite shield: A metal flange that fits over a foundation under the sill or around a pipe to act as a shield against the invasion of termites.

Thermal mass: A mass — concrete, masonry, water, or stone — designed to collect heat from the sun (or any other source) and release it when the heat source is removed and the ambient temperature drops.

Threshold: Wood or metal member tapered on both sides, used between door bottom and sill, and between jambs of interior doors, particularly when floors of different rooms are at different levels or of different materials. Thresholds are an integral part of an exterior door frame.

Tie beam: See Collar beam.

Timber: Lumber with width and thickness of at least 5 inches.

Timber support: A steel hanger for large beams.

Toenailing: Nailing at an angle, to connect one member with another member perpendicular to it.
Tongue: A bead of wood on the edge of a board cut to fit into the groove of another board.
Tongued and grooved: See Matched lumber.
Trap: A curved part of a plumbing drain that stays filled with water, to keep sewer gases from entering the house.
Tread: Horizontal board on a stairway; the part of the step that is stepped on.
Treenail: A wooden peg, used to hold posts and beams together in post and beam construction. The word is actually trunnel, corrupted from treenail.
Trim: Finish material on the interior and exterior of a house, not including interior walls and exterior siding. Also called woodwork.
Truss: Set of rafters, with collar beam and other members prebuilt and ready to install, connecting opposite wall points.

Undercoat: Primer or sealer for enamel in interior work.
Underlayment: A smooth material—plywood, particle board, or hardboard—installed on a subfloor as a base for finish material, such as carpeting, sheet flooring, or tile.

Valley: Inward angle formed when two sloping sides of a roof meet.
Valley rafter: The rafter that forms the valley.
Vapor barrier: Aluminum foil, kraft paper, or polyethylene designed to prevent passage of water vapor through or into exterior walls or floors. Always placed toward the heated part of the house. Used in conjunction with insulation and ventilation.
Varnish: A clear coating with a urethane or resin base.
Veneer: A thin ply of fine wood applied over a solid base in furniture, mainly to prevent warping.
Veneer plywood: Plywood made with several thin plies, as opposed to lumber-core plywood (which see).
Vent: Anything that allows air to flow through, as an inlet or outlet; a pipe to allow sewer gases to escape.
Ventilation: Any system that allows inflow and outflow of air.
Verge board: Wood board, usually fancy, covering the fly (end) rafter along the rake of a gable.

Wall: See Bearing wall; Nonbearing wall.
Wallboard: A catchall word for wall covering that is confusing because it means anything from plasterboard to composition board.
Wane: Bark or lack of wood on the edge of a corner of a board.
Warp: Any distortion of boards: crooking, bowing, cupping, or twisting, or any combination. Proper curing (drying) of wood can usually prevent or reduce warping. It often happens in service; that is, after boards have been installed, because moisture enters the wood, swelling it. It can often be prevented or minimized by thorough sealing, and by painting all faces, edges, and ends.

Glossary

Weatherstripping: Any material placed in window and door cracks to prevent passage of air. Made of bronze, wood with foam backing, or aluminum with vinyl tubing.

Weep hole: A small hole built into a wall, usually a retaining wall but often any masonry building wall, or in a fixed window frame, to drain water from one side to the other.

Whole-house cooling: A simple ventilating system designed to bring outside air into and through a building, dissipating it through the attic or other outlet high in the building.

Woodwork: See Trim.

Yard: A quantity of concrete, and sometimes of sand and gravel, that is actually a cubic yard, 27 cubic feet.

Further Reading

Anderson, L. O. *How to Build a Wood-Frame House*. New York: Dover Publications, 1973.
Brann, Donald R. *How to Build an Addition*. Briarcliff Manor, NY: Easi-Bild Directions Simplified, 1979.
———. *How to Build a Dormer*. Briarcliff Manor, NY: Easi-Bild Directions Simplified, 1972.
Clark, Sam. *Designing & Building Your Own House Your Own Way*. Boston: Houghton Mifflin Co., 1978.
Daniels, M. E. *How to Remodel & Enlarge Your Home*. Indianapolis/New York: Bobbs-Merrill, 1978.
De Cristoforo, R. J. *Housebuilding Illustrated*. New York/Evanston: Harper & Row, Popular Science, 1978.
Hotton, Peter. *So You Want to Build a House*. Boston: Little, Brown and Co., 1976.
———. *So You Want to Fix Up an Old House*. Boston: Little, Brown and Co., 1979.
Mazria, Edward. *The Passive Solar Energy Book*. Emmaus, PA: Rodale Press, 1979.
Philbin, Tom, and Fritz Koelbel. *How to Add a Room to Your House*. New York: Charles Scribner's Sons, 1980.
Shurcliff, W. A. *Air-to-Air Heat Exchangers for Houses*. Cambridge, MA: W. A. Shurcliff, 1981.
———. *Super Insulated Houses and Double Envelope Houses*. Andover, MA: Brick House Publishing Co., 1981.
———. *Thermal Shutters & Shades*. Andover, MA: Brick House Publishing Co., 1980.
Strickler, Darryl J. *Passive Solar Retrofit*. New York: Van Nostrand Reinhold Co., 1982.
Watson, Donald. *Designing & Building a Solar House*. Charlotte, VT: Garden Way Publishing, 1977.

Index

adhesive, 140–142
adhesive, construction, 132, 139, 140
adhesive, panel, 134
air conditioner, 123
air exchange, 204–206
anchor bolt, 31, 41, 42, 162
   ill., 31
antenna, 121, 123
apron, 111, 146, 164
   ill., 164
architect, 7–9
asphalt, 101, 104
attic, 18, 69, 102, 122, 123, 167, 219
   ill., 76, 79, 102, 212, 219

baluster, 168, 180, 181
balustrade, 181
   ill., 181
base, 179, 180
   ill., 180
baseboard, 131, 136, 137, 144, 150, 194, 196
   ill., 151, 196, 197, 199
basement, 7, 12, 14, 16, 26, 28, 29, 33, 36, 44, 121, 122, 124, 126, 160, 170–174, 192, 193
   ill., 14, 16, 33, 126, 171, 193
bathroom, 155, 156
bathtub, 125
   ill., 125
batten, 118
batten, reverse, 117
batter board, 28
beam, 20, 42, 44, 56, 68, 78, 80, 162, 166, 168, 182
   ill., 31, 32, 38, 54, 183
beam, collar, 64, 68
bids, 9
bird's mouth, 63, 65, 76
   ill., 77
blind-nailing, 116, 135, 137
block, 215
block, concrete, 34–36, 161, 164, 165
   ill., 34–36
block, corner, 146
   ill., 147
block, patio, 177, 185, 186
   ill., 186
Blueboard, 129, 133–134
   ill., 130
board and batten, 117
boiler, 174
   ill., 174
brace, 54, 57, 71, 74
   ill., 55, 75

bracket, 111
breezeway, 7, 160, 168–170
   ill., 161
brick, 175, 177–179, 185–187, 219
   ill., 108, 177–179, 186, 187
bridging, 47, 80
   ill., 48, 81

cable, 122
capital, 179, 180
   ill., 180
carpeting, 136, 139, 140, 150, 165, 170, 171, 216
carriage, 84, 86
   ill., 85
casing, 22, 53, 87–91, 93, 94, 113, 114, 136, 144, 146, 148, 149, 154, 159, 194, 201, 202
   ill., 22, 89, 90, 97, 119, 144, 146, 147, 148, 173, 201, 203
caulking, 89, 92, 114, 131, 132, 194, 201, 202
   ill., 203
cedar, red, 154
cedar, white, 154
ceiling, 18, 22, 36, 48, 60, 71, 81, 84, 122–124, 129, 131, 132, 150, 155, 158, 159, 165–168, 170–172, 191, 192, 193, 200–202
   ill., 16, 17, 60, 83, 126, 130, 151, 166, 192, 193, 199, 200
ceiling, cathedral, 19, 20, 68–70, 103, 160, 166–168, 170, 212
   ill., 20, 167, 212
cellulose, 10, 11, 19
cement, 29
cement, roofing, 105–108, 139, 164, 170, 171
   ill., 110
chimney, 68, 106, 108, 109, 217
   ill., 108, 109
clapboard, 98, 100, 112–116, 119, 162, 163
   ill., 113, 114, 120
cleat, 71, 76
collector, 209
column, 32, 35, 44, 179, 180
   ill., 32, 38, 45, 180, 181
compound, glazing, 154, 213, 214
compound, joint, 129, 131, 132, 200, 201
   ill., 130, 133
concrete, 25, 29–37, 41, 140, 142, 161, 165, 175, 178, 180, 181, 215
   ill., 28, 30, 32, 36, 37

contractor, 8, 9
cooling, 217–219
coping, 152
corner, 51, 54, 119–120, 132, 133, 142, 150, 152, 158, 159, 201, 202, 209
   ill., 17, 35, 52, 85, 120, 134, 141, 158, 177, 203, 211
corner bead, 133
cornerboard, 95, 119, 120
cornice, 95, 96, 98
   ill., 96
counterflashing, 108
   ill., 108
course, 34–35, 104, 105, 107, 109, 113–116, 120, 162, 163, 178
   ill., 34, 104, 105, 117, 178
crawl space, 12, 14, 16, 35–38, 41, 44, 121, 124, 126, 162
   ill., 15, 16, 36, 38, 126, 162, 193
curbing, 185
cypress, 154

dead bolt, 150
deck, 155, 157, 182
   ill., 182, 184
door, 22, 51, 53, 55, 60, 87, 92–94, 113–116, 118, 144, 146, 148–150, 155, 159, 161, 163–165, 172, 201, 219
   ill., 53, 92–94, 148, 149, 158, 173, 202, 218
door, exterior, 92
   ill., 92
door, French, 212
   ill., 6
door, screen, 94
door, sliding, 212
door, storm, 94, 174, 177
doorknob, 148
doorway, 149
dormer, 3, 5, 67, 71, 72–80
   ill., 5, 73, 74, 77, 78, 80
dormer, shed, 78, 100
   ill., 5, 72, 73
downspout, 111
drain, 32, 124, 125
drip cap, 89
   ill., 89
drop edge, 68, 96, 100, 103, 104, 109, 114, 116, 162, 178
   ill., 97, 103, 104, 107
driveway, 161, 163, 164
   ill., 164
dry wall, 129, 134, 170, 171, 201

238

Index

dry well, 33, 111, 163, 198
    ill., 164
duct, 121, 123, 125, 126, 174, 193, 196, 205, 209
    ill., 128, 174

eave, 64, 68, 73, 95, 99, 100, 102–104, 110, 113, 166, 192
    ill., 63, 96, 102–104
electricity, 121
ell, 3, 25
    ill., 4
enamel, 155, 156
excavation, 28–29
exchanger, heat, 204–206
    ill., 205

face-nailing, 64, 67, 76, 78, 105, 114, 116, 136, 137
facia, 68, 95, 96, 100, 103
    ill., 77, 96, 97, 99
fan, ceiling, 217, 219
    ill., 218
felt, 101–104, 110, 111
felt, roofing, 30, 33, 74, 88, 89, 93, 98, 113, 114, 118–120, 136, 142, 162, 187
fiberglass, 10, 11, 15, 16–19, 33, 49, 54, 55, 81, 104, 165–168, 171, 191, 192, 194, 195, 198, 199
    ill., 15–17, 103, 195, 199
fir, 157
fireplace, 215
fire stop, 131, 165
    ill., 165
flashing, 42, 69, 100, 106–110, 163, 180
    ill., 100, 106, 107, 163
flashing, rolled, 107
flashing, step, 107, 108
    ill., 108
floor, 16, 17, 33, 35, 36, 39, 41, 42–48, 51, 53, 58, 71, 73, 76, 78, 80, 82–84, 93, 94, 123–126, 136–142, 148, 155, 157, 158, 161, 162, 165, 166, 168, 170–171, 182, 193, 197, 201, 202, 216, 219
    ill., 16, 17, 31, 34, 81, 84, 122, 151, 162, 164, 171, 218, 219
floor, attic, 18, 22, 191, 219
    ill., 18, 19, 192
floor, concrete, 215
    ill., 139
floor, finish, 150
floor, second, 53, 56, 57, 60, 71, 72, 81, 82, 166, 167
    ill., 6, 59, 81
floor covering, 142
flooring, finish, 136–143
    ill., 137, 138, 139
footing, 28–30, 32–35, 37, 39, 41, 161, 175
    ill., 28, 29, 32, 161, 176
form, 29, 30
    ill., 176
form boards, ill., 29
foundation, 14, 25, 27, 29–39, 41, 42, 44, 57, 58, 116, 121, 160–163, 168, 170–172, 175
    ill., 14, 15, 27–29, 31, 33, 36, 37, 39, 40, 45, 50, 58, 161, 162, 164
frieze board, 95, 96, 98, 113, 116
frost line, 29, 39, 161
furnace, 174
    ill., 174
furring strip, 129, 134, 135
    ill., 130

gable, 64–66, 68, 80, 95, 98, 100, 116, 117, 168
    ill., 65, 69, 72, 80, 98, 100, 169, 179
gable, half, 5, 78, 80, 116
    ill., 5, 78, 79
garage, 5, 7, 160–170
    ill., 6, 161, 162, 164, 166
girder, 30–32, 36–38, 42, 44, 46
    ill., 31, 37, 38, 45
glazing, 21, 155, 185, 216
gravel, 30, 33, 39, 111, 175
greenhouse, 13, 209, 213
    ill., 211
grout, 142, 143
guide board, ill., 116, 117
gutter, 63, 111
    ill., 111

hanger, 111
hanger, joist, 46, 60, 65, 84
hanger, strap, 111
hardwood, 157
header, 53–55, 60, 64, 73, 84, 125, 198
    ill., 47, 53–55, 57, 198
heat, auxiliary, 217
heat, intrinsic, 191
heat, solar, 12–13, 41, 63, 80, 87, 95, 168, 170, 209–214, 215, 216
    ill., 13, 169, 210, 211, 216
heating, 125–128, 168, 170, 191, 193, 204
hinge, 148, 149
    ill., 149
hip, 67, 68, 105, 110

ice dam, 102, 110
    ill., 102
insulation, 10, 11, 14, 16–18, 20–22, 32, 33, 36, 39, 41, 42, 48, 49, 51, 54, 58, 68, 70, 72, 73, 80, 81, 93, 102, 121–124, 126, 129, 139, 153, 162, 165–167, 170–172, 191–203, 217; *see also* superinsulation
    ill., 14–20, 32, 33, 54, 101, 102, 122, 123, 127, 128, 167, 171, 172, 192, 193, 195–198, 200
insulation, duct, 128
    ill., 128

jamb, 53, 87, 88, 90, 91, 93, 144, 146, 148, 150, 172, 194, 201, 202
    ill., 90, 93, 144, 146–148, 201
jamb extender, 93
    ill., 93
joist, 18, 19, 22, 28, 35, 38, 42, 44, 46–49, 51, 54, 56–58, 60, 62, 64, 68, 70, 71, 74, 76, 80, 81, 84, 122, 124–126, 129, 131, 140, 161, 162, 164–168, 172, 174, 182, 191–193, 199, 200
    ill., 16, 18, 19, 31, 37, 45, 47, 50, 59, 60, 66, 76, 81, 84, 127, 130, 162, 164–168, 171, 172, 183, 184, 192, 193, 199, 200
junction box, 124

key, 30
kitchen, 155, 156
knot, 154, 156

ladder, ill., 168
landing, 86
    ill., 85
latch, 150
lath, 114, 194
lattice, 38
ledger, 44–46, 65, 96, 182
    ill., 45, 97, 183
light, 123
lock, 150
loft, 167, 168
    ill., 168, 169
lookout, 65
    ill., 65

239

Index

masonry, 170, 185, 212, 215, 217
   ill., 216
mass, thermal, 212, 215, 217
mesh, 35, 39, 41, 133, 175
   ill., 130, 176
mineral wool, 10, 11, 18
miter, 152
   ill., 152
moisture, 162
molding, 100, 135, 144, 150, 152, 194
   ill., 145, 151, 152
molding, band, 119, 150
   ill., 119, 150
molding, base, ill., 150
mortar, 34, 35, 142, 143, 175, 177, 178, 187
mortise, 149, 150
   ill., 149
mover, house, 161

nailer, 60, 100
   ill., 60
nosing, 83
   ill., 83

opening, rough, 51, 53, 88, 93, 146, 148
   ill., 90
outlet box, 123, 124, 131
   ill., 123, 132
overhang, 13, 18, 58, 62–65, 68, 70, 81, 90, 95, 96, 98, 100, 116
   ill., 13, 19, 58, 59, 63, 69, 98, 99
overlap, 162

paint, 153–154, 155–156
paint, ceiling, 155, 157
paint, cement-based, 170
paint, latex, 153–156
paint, oil, 153–155, 157
paint, texture, 155
   ill., 156
paint, vapor-barrier, 191, 192, 194
   ill., 192
paneling, 60, 132, 134, 135, 170, 172
   ill., 135
parquet, 136, 138
   ill., 139
parquet tiles, 136, 142
partition, 56, 58, 60, 123, 131, 170
   ill., 56, 57, 60, 130
paste, 157, 158
patio, 175, 185–187
   ill., 6, 186, 187
paving, ill., 177

permits, 7, 8
pier, 35–38, 182
   ill., 37, 38, 182, 183
pilaster, 180
pillar, 179, 180
   ill., 180, 181
pine, 157
pipe, 121, 122, 124, 125, 193, 197
   ill., 16, 121, 122, 126, 197
pipe, vent, 109, 124
pitch, 62, 63, 66, 72, 95, 110
plank, 136, 138
plans, 9
plaster, 129, 133, 157, 170, 192, 195
   ill., 130, 195, 200
plasterboard, 22, 48, 49, 53, 60, 86, 90, 94, 129, 131, 132, 134, 157, 172, 192, 194, 197, 198, 200
   ill., 130, 133, 200
plate, 165
plate, bottom, 166, 199
plate, floor, 33, 51, 53–55, 76, 78, 80
   ill., 124, 164
plate, top, 33, 51, 54, 55, 57, 61, 62, 64, 68, 78, 80, 165, 166, 171
   ill., 60, 167, 171
platform construction, 42
   ill., 43
plenum, 126
plumbing, 121, 124–125, 160
   ill., 124
pocket, 31, 44
   ill., 31
polyethylene, 11, 41, 48, 165, 170
   ill., 14, 32
polyisocyanurate, 10, 11, 17, 20, 197
porch, 160, 175, 177, 181, 182, 185
   ill., 179, 181, 182
porch, solar, 213
post, 71–73, 78, 86, 168, 182
   ill., 32, 85, 182–184
preservative, 154, 155
preservative, wood, 95, 180
primer, 95, 154
primer, metal, 155
putty, 154–156

quarter-round, 119, 150

radiator, 126, 195, 197, 198, 217
   ill., 196–199
rafter, 19, 56, 57, 62–68, 71, 73, 74, 76, 78, 80, 88, 95, 96, 99, 100, 166, 167, 168, 192

   ill., 20, 67, 73, 75–77, 79, 96, 97, 99, 166, 167
rail, 168, 181
railing, 168, 169, 180, 182
   ill., 184
rake, 64, 65, 68, 95, 96, 98, 99, 103, 104, 110, 116
   ill., 65, 98, 99, 103
redwood, 154
reflector, 196
   ill., 196
register, 126
reinforcing, 29, 34, 164
   ill., 50
return, cold-air, 126
   ill., 127
return, eave, 98, 99
   ill., 99
R factor, 11, 19, 20–22, 32, 49, 167, 192, 194, 197, 198
ridge, 66, 68, 72–74, 78, 105, 110, 164
   ill., 64, 78, 105
ridgebeam, 80
   ill., 79
ridgeboard, 61, 63, 64, 66–68, 76, 78, 99
   ill., 67, 73, 76, 77, 79
rigger, 161
rise, 72, 74
riser, 83, 84, 175
   ill., 83, 85
riser, duct, ill., 218
Rocklath, 129, 194
rod, 175
   ill., 176
roof, 20, 53, 57, 61–70, 71, 72, 74, 80, 95, 96, 98, 99, 101, 102, 106–109, 160, 167, 168, 177, 179–181, 212
   ill., 19, 20, 62, 63, 65, 67, 73, 74, 97–99, 103, 106, 108–110, 179
roof, built-up, 101, 110
roof, flat, 69, 70, 101, 110
roof, gable, 61, 67
   ill., 4, 62, 67
roof, gambrel, 66, 67
   ill., 67
roof, hip, 66, 69, 96
   ill., 66, 69
roof, shed, 65, 100, 106
   ill., 4, 62, 66, 107
roofing, 65, 101–111
roofing, roll, 72, 101, 102, 103, 106,

Index

109, 110
   ill., 102, 104, 110
roofline, ill., 111
rug, 140
run, 61–63, 72
rust, 155

sash, 87, 89, 90, 92, 154
saddle, 109
   ill., 109
scaffolding, 71, 72
screed, ill., 30
screen, 92, 177, 182, 185
   ill., 184, 185
scroll, ill., 156
seam, rolled, ill., 107
seam, standing, 106
   ill., 106
seepage, 170
sewer, storm, 111
sheathing, 20, 42, 44, 49, 54, 57, 58, 60, 64, 65, 68–70, 74, 76, 80, 81, 87–90, 96, 98, 100, 102, 103, 105, 107, 111, 118, 162, 199
   ill., 15, 20, 57, 74, 98, 99, 116, 167, 199
sheathing paper, 112, 136, 180
shellac, 95, 154, 156
shim, 46, 89, 93, 129, 149
shimming, 116
shingle, 89, 98, 100–103, 108, 114–116, 119, 154, 162, 163
   ill., 104–106, 113, 115, 117, 120, 163
shingle, asphalt, 101
   ill., 103
shingle, fiberglass, 104
shingle, roof, 20, 68, 69, 72, 74, 107, 109
   ill., 20, 107, 108
shoe mold, 137, 142, 150
   ill., 151
shutter, 21, 22, 91, 174, 201, 212
   ill., 21, 22, 91
sidelight, 92, 93
   ill., 94
siding, 44, 49, 57, 60, 65, 87, 88, 100, 107, 108, 112–120, 154, 162, 163, 199
   ill., 15, 113, 162, 199
siding, vertical, 117–119
   ill., 117, 118, 119
sill, 17, 30, 31, 33, 35, 41, 42, 44, 45–47, 57, 58, 89, 93, 113, 116, 154, 155, 161, 164, 171

ill., 16, 31, 44, 45, 58, 90, 162, 172
sill sealer, 31, 35, 42
size, glue, 157
skimcoat, ill., 130
skirtboard, ill., 38
skylight, 13, 109, 168, 209, 212
   ill., 169, 210, 212
slab, 25, 32, 39, 41, 121, 139, 160, 163, 170, 175, 177, 204, 215
   ill., 14, 32, 40, 121, 176, 177
slate puller, 163
   ill., 163
sleeper, 20, 139, 171
   ill., 20, 60, 130
soffit, 65, 95, 96, 98–100
   ill., 77, 96, 97, 99
solar heating. *See* heat, solar
Sonotube, 35–36, 37
   ill., 37
sound, 22
span, 61, 63
square, framing, 63
   ill., 64
staircase, 82–86, 167, 168
   ill., 82, 83, 85
stairway, 122, 123
   ill., 83, 84
stain, 139, 153, 154–155, 156
stain, semitransparent, 154, 155
stain, solid, 154
step, 82, 83, 86, 175, 177
   ill., 85, 176, 186
step, brick, 177, 178
   ill., 178, 179
stone, solar-collecting, 215–216
stone dust, 186, 187
stool, 144, 146
stoop, 175–179
   ill., 176
stop, 90, 91, 146, 148, 149, 201
   ill., 93, 146–148, 201
storage, heat, 212
story, second, 5, 58, 72, 80–81, 83
stove, 217
   ill., 218
straightedge, 35, 45, 46, 55, 65, 159
strapping, 170, 171
   ill., 171
striker, 150
string, 27
   ill., 27
strip flooring, 136–139, 142, 165, 170
   ill., 137–139
stud, 17, 22, 42, 49, 51, 53–55, 58,

64, 76, 78, 86, 89, 90, 93, 96, 122, 132, 146, 165, 166, 171, 182, 185, 194, 195, 198, 199, 202
   ill., 15, 17, 50, 55, 56, 65, 67, 124, 125, 128, 132, 165, 198, 199
Styrofoam, 10, 11, 14, 15, 32, 33, 35, 39, 41, 42
   ill., 14, 15, 32, 33
subfloor, 47, 48, 51, 54, 72, 83, 136, 140, 165
sump, 30, 33, 170
sump pump, 33, 170
sun space, 13, 209, 213
   ill., 12, 211
superinsulation, 10–11, 14–23, 48, 49, 68, 74, 76, 90, 121, 165, 167, 170, 174, 191–203, 217, 219; *see also* insulation
   ill., 75, 77
switch box, 123, 124, 131

tape, 132, 200, 201
   ill., 130, 133, 134
tape duct, 167
tar, 33, 110, 111, 139, 170, 171
   ill., 139
termite shield, 31, 35, 42
   ill., 44
threshold, 92, 143, 175
tie, collar, 166
   ill., 166, 167
tile, 48, 136, 139, 140–143, 165, 170, 215
   ill., 141, 142
tile, drain, 30
timber, landscape, 185, 186
toenailing, 46, 53, 56, 58, 64, 65, 71, 74, 76, 96, 98, 165, 180
   ill., 46, 47
tread, 82, 83, 175, 178
   ill., 83
trim, 95–100, 135, 144–152, 155, 156, 163, 172, 194
   ill., 96, 97, 98, 99
trunk line, 126
   ill., 127
Tyvek, 113

undercoat, 155
undercoater, 155, 156
underlayment, 48, 136, 140, 142, 143, 150, 165, 170
urethane, 11

Index

vacuum system, 121, 123
valley, 67–69, 106, 133
   ill., 66, 106, 133
vapor barrier, 10–12, 14, 16–18, 41, 48, 49, 123, 132, 139, 162, 165, 168, 171, 172, 191–194, 197, 198, 215
   ill., 17, 32, 123, 139, 172, 193, 195
varnish, 155–157
vent, 196, 198, 213, 214, 219
   ill., 196, 219
vent, eave, 70
vent, ridge, 19, 20, 69, 70, 105, 166
   ill., 19, 69
vent, roof, 109
vent, soffit, 19, 20, 166
   ill., 19, 69
ventilation, 12, 19, 36, 63, 69–70, 80, 96, 102, 155, 204
   ill., 36, 96, 101, 102
ventilator, ill., 69

wainscoting, 135, 170
wall, 27, 39, 49–60, 94, 106, 107, 123–126, 129–135, 142, 155, 157–159, 163, 165, 166, 170–172, 174, 194–203, 204, 216
   ill., 15, 17, 50, 52, 55, 56, 59, 60, 73, 75, 76, 79, 80, 107, 108, 113, 123, 132, 147, 151, 164, 166, 167, 169, 171, 173, 195–199, 201–203, 211
wall, basement, 193
wall, double, 17, 42, 49, 53–56, 121, 123, 124, 126, 201
   ill., 17, 43, 50, 90, 122, 128, 198, 201, 202
wall, knee, 76, 78
   ill., 77, 79
wall, stud, 33, 74
   ill., 51
wall, trombé, 214
wall, window, 212
wallpaper, 134, 157–159, 194
   ill., 158
water, 215
water table, 163
   ill., 163
water vapor, 11, 20, 153, 155, 170, 171, 192, 204, 215
   ill., 195
weatherstrip, 201
weight, 201
window, 13, 21, 51, 53, 55, 60, 80, 87–92, 94–96, 113–116, 118, 119, 144, 159, 165, 166, 168, 172, 174, 195, 201, 202, 204, 209, 212, 213, 219
   ill., 53, 78, 79, 88, 114, 169, 173, 198, 206, 210, 213
window, awning, 87
   ill., 79, 88
window, bay, 87
   ill., 88
window, casement, 87, 90
   ill., 88
window, clerestory, 80, 87
   ill., 78
window, double-hung, 87
   ill., 88
window, hopper, 87
   ill., 79, 88
window, picture, 87
window, roof, 109, 209, 212